The role of crack growth
in metal fatigue

Many accidents on railways are to be ascribed
to that progressive action which may be termed
'the fatigue of metals'.

F. Braithwaite,
*Proceedings of the Institution
of Civil Engineers*,
1854, xiii, 463

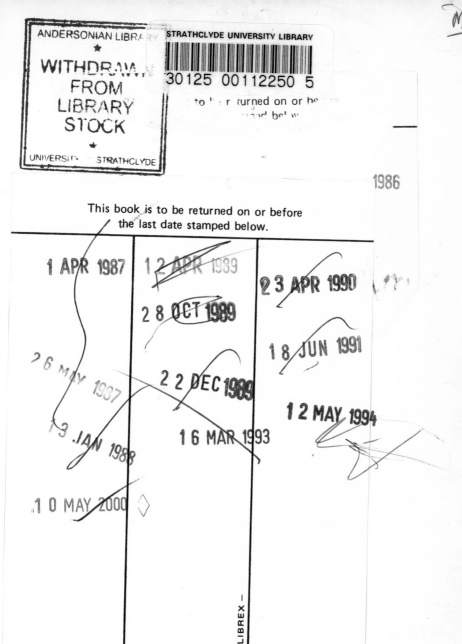

The role of crack growth in metal fatigue

L. P. POOK

*National Engineering Laboratory
East Kilbride, Glasgow*

The Metals Society
London

Book 307 published by

The Metals Society
1 Carlton House Terrace
London SW1Y 5DB

ISBN 0 904357 63 5

Text and titles set in Linotron Imprint

Printed and made in England
by J. W. Arrowsmith Ltd, Bristol

Contents

Acknowledgments

This book is published with the permission of the Director of the National Engineering Laboratory of the Department of Industry. The author acknowledges gratefully the assistance, in a variety of ways, of many of his colleagues at NEL.

Preface

It's all a hoax, but it pays the bills.
My wife.

This book has its origins in extensive discussions with Mr N. E. Frost and Dr K. J. Marsh about the form a successor to our book *Metal fatigue* (1974, Clarendon Press) might take and incorporates many of their ideas, but the plan eventually adopted is entirely my own. The first part of this short book takes a wide view of the subject of metal fatigue and its development. This is followed by a short chapter on the basis of design against metal fatigue. The remainder takes a more detailed look at selected topics, mostly concerned with fatigue crack growth, where experience has shown that misunderstandings are common. Discussion with colleagues at the National Engineering Laboratory and elsewhere confirmed my idea that there was a need for such a book, so I make no apology for adding to the already voluminous literature. The final form of the book has been influenced by my family's tolerant scepticism, which has, I hope, prevented me from taking myself and my views too seriously.

Notation

Only those symbols which appear in more than one section are listed

a	Crack length
a_f	Final crack length
a_0	Initial crack length
δa	Increase in crack length
D	Constant in equation (4.4), fractal dimension
E	Young's modulus
K	Stress intensity factor, subscripts I, II, III denote mode
K_c	Critical value of K_I for crack growth
K_{Imax}	Maximum value of K_I in fatigue cycle
K_{Imin}	Minimum value of K_I in fatigue cycle
ΔK_{eff}	Effective value of ΔK_I
ΔK_{th}	Threshold value of ΔK_I
ΔK_I	Range of K_I in fatigue cycle
m	Exponent in equation (4.4); constant in equation (6.4)
N	Number of cycles
N_i	Number of cycles to cause failure at ith load level
n_i	Number of cycles at ith load level
R	Stress ratio (S_{min}/S_{max})
r, θ	Polar coordinates (Fig. 4.3)
r_p	Radius of plastic zone at crack tip
S	Stress
S_a	Alternating stress
S_m	Mean stress
S_{max}	Maximum stress in fatigue cycle
S_{min}	Minimum stress in fatigue cycle
S_Y	Yield stress
T	Temperature, time, stress parallel to crack
x, y, z	Cartesian coordinates (Fig. 4.3)
α	Geometric correction factor in equation (4.2)
ν	Poisson's ratio
ρ	Notch root radius
σ	Root mean square value of S, standard deviation
$\sigma_x, \sigma_y, \sigma_z$	Direct stress components (Fig. 4.3)
$\tau_{xy}, \tau_{yz}, \tau_{xz}$	Shear stress components (Fig. 4.3)

CHAPTER 1

Introduction

Fate is not satisfied with inflicting one calamity.
Publilius Syrus, *Sententiae*, 274.

An early dictionary definition of metal fatigue is[1] 'The condition of weakness in metal caused by repeated blows or long-continued strain'. Rigorous definition of metal fatigue is difficult[2] and a satisfactory modern dictionary definition does not appear to exist. The following is offered: 'failure of a metal under a repeated or otherwise varying load which never reaches a level sufficient to cause failure in a single application'.

To the outsider, metal fatigue presents a puzzling picture. Since 1838 it has amassed a great body of literature[3]—over 20 000 papers on fatigue have now been published[4] throughout the world—yet experts on metal fatigue often appear incapable of giving straightforward answers to what enquirers regard as simple questions. In a sense, the problem appears to have been largely solved in that catastrophic failures due to fatigue in any type of engineering structure are rare.[5] Official inquiries into catastrophic fatigue failures, involving loss of life or major financial loss, usually identify a clear reason for the failure and often indicate apparent human negligence: a recent example is the loss of the accommodation rig *Alexander L. Kielland*.[6] Lesser fatigue failures, often unrecognized,[7] are common and cause a great deal of inconvenience and expense. Figures 1.1–1.4 show a selection of such nuisance failures that occur in everyday life. It is easy to imagine circumstances in which some could have had serious consequences, with the underlying cause perhaps going unrecognized. Beyond advice on the need to reduce stress concentrations, there often appears to be little guidance on how to avoid fatigue failures. Indeed, in the past most advances seem to have been made on a trial-and-error basis.[8] Nowadays, it is becoming more generally accepted in general mechanical engineering circles that fatigue testing of prototype structures or components under realistic loading conditions is often an important part of the design and development process.[9,10]

1

Fig. 1.1   Fatigue failure of driving dogs on a motorcycle gear [×2]

It is hoped that this book will assist non-specialists in understanding the literature on metal fatigue, and that a reasonable balance has been found between the conflicting requirements of readability and academic rigour. In particular, it attempts to explain why a subject which has received so much attention should appear to be in such

Fig. 1.2 Fatigue failure of the diaphragm in the control gear of a domestic freezer [×5]

Fig. 1.3 Fatigue failure of a hammer head

an unsatisfactory state. It is assumed that readers are familiar with strength of materials theory and elementary metallurgy, and have some acquaintance with the literature on metal fatigue.

This book does not aim to be comprehensive. The effect of corrosive environments is scarcely mentioned, despite their practical importance. There is little of metallurgical interest, for at room temperature cyclic stressing usually causes little significant metallurgical change. Nor is there any discussion of interactions between

Fig. 1.4 Fatigue failure of the steel reinforcement of a lady's shoe

creep and fatigue at elevated temperatures. Nevertheless, it is believed that this book treats the core of the subject which must be understood before detailed study of particular topics is undertaken. Newcomers often find metal fatigue a vague and difficult subject, and particular attention is paid to aspects which experience has shown to be frequently misunderstood. Detail which is not essential to the arguments used is omitted, especially mathematical detail which too often obscures rather than illuminates. Where not referenced, such detail may be found in any of the modern books on metal fatigue.[11-16]

The next chapter is an historical survey of the development of ideas on metal fatigue, and Chapter 3 reviews the basis of design against metal fatigue. Some discussion of legal aspects is included as these are becoming increasingly important under present and impending legislation. Developments in linear elastic fracture mechanics have made it possible to treat many aspects of fatigue crack growth quite rigorously: basic ideas are introduced in Chapter 4, and some more advanced topics are covered in Chapters 6 and 7 which may be omitted on a first reading. Some statistical aspects of fatigue are discussed in Chapter 5. The book closes with a chapter that draws together the main points made in a summary of the present position.

SI units are used throughout, apart from an example given in Section 3.3 which is based on a standard written in Imperial units. All symbols used are defined the first time they appear. Those that are used in more than one section are collected in the *Notation*. Where possible, symbols conform to common usage, but numerous variations occur in the literature, and it was not possible to avoid some inconsistencies.

REFERENCES

1   J. A. H. MURRAY (ed.): 'A new English dictionary on historical principles, Vol. 4', 1901, Oxford, Clarendon Press
2   L. P. POOK: *J. Soc. Environ. Eng.*, 1976, **15-4**, (71), 3–10
3   J. Y. MANN: 'Bibliography on the fatigue of materials, components and structures, Vol. I, 1838–1950', 1971; 'Vol. II, 1951–1960', 1978; Oxford, Pergamon Press
4   T. YOKOBORI: 'On the critical problems in physico-mechano-structural foundations of fracture', in 'Advances in fracture research, Vol. 3' (ed. D. Francρis), 1145–1166, 1981, Oxford, Pergamon Press
5   A. G. PUGSLEY: 'The safety of structures', 1966, London, Edward Arnold
6   NORWEGIAN PUBLIC REPORT: 'The *Alexander L. Kielland* accident' (in Norwegian with extended summary in English), 1981, Oslo, Bergen and Tronsø, Universitetsforlaget
7   O. E. LISSNER: *J. S. Afr. Inst. Min. Metall.*, 1967, **67**, (6), 273–316
8   R. R. WHYTE: 'Engineering progress through trouble', 1975 London, Institution of Mechanical Engineers
9   K. J. MARSH: *J. Soc. Environ. Eng.*, 1974, **13-4**, (63), 15–16, 21–22
10 K. J. MARSH: *Int. J. Fatigue*, 1979, **1**, (1), 3–6

11 N. E. FROST, K. J. MARSH and L. P. POOK: 'Metal fatigue' 1974, Oxford, Clarendon Press

12 T. R. GURNEY: 'Fatigue of welded structures', 2nd ed., 1979, London, Cambridge University Press

13 S. KOCAŃDA: 'Fatigue failure of metals', 1978, Alphen aan der Rijn, Sijthoff and Noordhoff International Publishers

14 H. O. FUCHS and R. I. STEPHENS: 'Metal fatigue in engineering', 1980, New York, Wiley–Interscience

15 C. BATHIAS and J.-P. BAÏLON (eds.): 'Le fatigue des matériaux et des structures' ('Fatigue of materials and structures'), 1980, Paris, Maloine SA

16 M. KLESNIL and P. LUKÁŠ: 'Fatigue of metallic materials', 1980, Amsterdam, Elsevier Scientific Publishing Co.

CHAPTER 2

Historical survey

His reign is marked by the rare advantage of furnishing
very few materials for history; which is, indeed, little
more than the register of the crimes, follies, and mis-
fortunes of mankind.

Gibbon, *Decline and Fall of the Roman Empire*, ch. 3.

2.1 INTRODUCTION

This chapter is a short survey up to the present day of fatigue, its
development and how it interacts with design. It is intended to set
the scene for the more specialized topics developed in later chapters.
The first section briefly reviews the literature of metal fatigue as
exemplified by books on the subject and related topics. The remaining
three sections cover mechanisms of fatigue, cumulative damage and
fatigue testing.

Metal fatigue is largely a descriptive subject, which accounts for
its extensive literature.[1] The descriptions used can be divided into
two broad groups, metallurgical and mechanical. Metallurgical
descriptions are concerned with the state of the material before,
during and after the application of cyclic loads, and are usually taken
to include the study of mechanisms. Mechanical descriptions are
concerned with the mechanical response to a given set of loading
conditions, for example, the number of cycles to failure or the rate
of growth of a fatigue crack. Mechanical descriptions are the more
useful from an engineering viewpoint, where service behaviour must
be predicted, and are therefore given more emphasis in this book.
The basic analytic framework for the description of fatigue is provided by the branch of applied mechanics known as strength of
materials. As much fatigue work is concerned with crack growth,
fracture mechanics, which is the applied mechanics of crack growth,
is also used.

2.2 BRIEF REVIEW OF SOME BOOKS ON METAL
FATIGUE AND RELATED SUBJECTS

Although numerous books on metal fatigue have been written, nothing approaching a 'standard text' has emerged. There is no equivalent

to Timoshenko[2] or Roark[3] in strength of materials, or Cottrell[4] or Newton[5] in metallurgy. Perhaps the nearest are a recent French book[6] and, from the designer's viewpoint, a recent American book.[7]

Two classic books, both entitled *The fatigue of metals*, were published in the mid-twenties: one, by Gough,[8] was published in 1924, and the other, by Moore and Kommers,[9] in 1927. Both books contain detailed descriptions of fatigue-testing machines, methods of determining the fatigue limit at zero mean load, and tests aimed at deriving so-called fundamental theories of fatigue, usually from bulk specimen properties. The data given in both books were all obtained from simple specimens.

A number of similarly titled volumes followed in the 1950s and '60s.[10-13] These again emphasize the data obtained from simple specimens, especially notched specimens, but none contain any significant fatigue crack growth data. Although not strictly a book, the proceedings of the Institution of Mechanical Engineers' International Conference on Fatigue,[14] held in 1956, provides a convenient description of the existing state of the art.

Worthwhile descriptions of fatigue crack growth started to appear in books in the 1970s; *Metal fatigue* by Frost, Marsh and Pook, published in 1974[15] was the first to utilize a fracture mechanics approach, based on stress intensity factors, wherever possible. More recent books[6,7,16,17] are essentially refinements of their approach. Plastic flow is an essential feature of the mechanisms of both crack initiation and crack growth. Plastic flow depends upon the movement of dislocations, which are regions of imperfection in the crystal lattice. Most books on fatigue therefore contain elementary descriptions of dislocations, and various detailed accounts are available, for example that by Cottrell.[18] Introductory books on metallurgy usually contain a section on fatigue in which the basic principles are enunciated.[4,5] The brief description given by Newton in 1947[5] reads as if it were written much later. A popular book on metals[19] published in 1951 contains a good introduction to metal fatigue, but this has not been updated for the current edition.[20] An otherwise excellent popular book[21] on structural behaviour virtually ignores the topic.

The implications of early work on fatigue were quickly taken up in design procedures; for example, Unwin's book on machine design,[22] published in 1894, deals briefly but, for its time, adequately with the effects of cyclic loading. A section on fatigue is often included in engineering handbooks, but these are usually not very helpful. Undergraduate books on engineering materials usually contain a section on fatigue, for example Pascoe[23] gives a straightforward elementary introduction, but little guidance on how to design against fatigue. Some books on fatigue promise assistance with design against fatigue.[24-26] These suffer from the drawback that they are essentially collections of examples where particular approaches have proved successful and therefore offer little help if the problem under consideration is not similar to one of the examples. In practice, the most satisfactory guidance to design against fatigue is found in standards, such as BS 5400 for bridges.[27] In their book[7] *Metal fatigue in engineering*, Fuchs and Stephens give a full description of basic principles as well as a wide range of examples. The general questions of how the safety of structures can be assured, and what should be regarded as acceptable levels of safety, are discussed by Pugsley.[28]

More specialist books started to appear in the 1960s. Forsyth's book[29] examines the subject from the viewpoint of a metallurgist, and describes fractographic techniques for the examination of fatigue fracture surfaces. Statistical treatments appropriate to fatigue test results are discussed by Johnson.[30] A book by Harris[31] concentrates on fatigue of aircraft structures, and one by Sandor[32] on low cycle fatigue and associated specimen behaviour. *Fatigue of welded structures* by Gurney[33] appeared in 1968. The recently published second edition,[34] incorporating fracture mechanics where appropriate and much useful advice on design, may well become a standard text. The American Society for Testing and Materials has recently published a handbook on fatigue testing.[35] A book on the statistical design of fatigue experiments[36] was published in 1975.

During the 1970s, books on fracture mechanics started to appear. There is considerable overlap between topics regarded as belonging to metal fatigue and to fracture mechanics, and four volumes[37-40] include extensive discussion of fatigue crack growth. Another[41] includes much information on the application of fracture mechanics to design against fatigue, but again this is essentially a collection of examples.

Metal fatigue receives little mention in fiction. Neville Shute's *No highway*[42] is the only well known novel where it forms a central element in the plot. Perhaps less well known is Shute's expertise in engineering.[43]

2.3 DEVELOPMENT OF MODERN IDEAS ON THE MECHANISMS OF FATIGUE CRACK INITIATION AND GROWTH

About a hundred years ago discussions were taking place at the Institution of Mechanical Engineers[44] on the fracture of wrought-iron railway axles, one member stating that he personally knew of thousands of such breakages. It was generally accepted that the failures were a consequence of the cyclic nature of the loading, and all agreed that the fracture faces of broken axles had a crystalline appearance. This convinced many members that cyclic stressing could make tough fibrous iron become crystalline and brittle. Others, however, were not so convinced and this inevitably led to much technical argument regarding the mechanism by which cyclic stressing resulted in a macroscopically brittle failure. This interesting but somewhat sterile argument was finally settled when the optical microscope was used to study metallurgical structures and fatigue processes.

It was known in the 1920s[8,9] that fatigue failure is a progressive and localized process involving both the initiation of a crack and its progressive spreading, and it was appreciated much earlier that cracks could spread slowly in service.[44,45] Unfortunately, there was widespread failure to appreciate the fact that fatigue consists of crack initiation and growth, and this led to considerable confusion over the nature of fatigue.[46] Indeed, in the 1920s and 1930s engineers in general felt that fatigue was somehow a balance between damage and repair. The following anonymous description, quoted by Frost,[46] in its 1920s/30s physical-metallurgical phraseology, aptly sums up the position:

> Fatigue can be regarded on the one hand as a destructive action bringing about dislocation and progressive cracking of the crystals and on the other as an ameliorating action arising from the phenomenon of accommodation. Fatigue resistance is conditioned by the resultant of these two actions, the point of equilibrium being the fatigue limit.

Many theories of fatigue were advanced; all failed because they tried to explain too much,[46] i.e. both the initiation and the growth (albeit often unrealized) of a crack by cyclic stressing. For example, phenomena such as the behaviour of grey cast iron can be explained simply once it is realized that they are the responses of a cracked body; explanations and arguments become involved and illogical when attempts are made to explain the same phenomena in terms of an uncracked body. Although information on the mechanisms of fatigue is freely available, confusion persists to this day; the following statement appears in a current engineering handbook: 'Fatigue is generally understood as the gradual deterioration of a material which is subjected to repeated loads'.

The first work on fatigue crack initiation was done around 1900–1910 by Ewing and Humphrey.[47] They prepared the surface of a specimen metallurgically and examined the surface grains with an optical microscope during the course of a fatigue test. They observed the formation of slip lines across grains, their broadening into bands and the eventual development of cracks in the broadened bands. Half a century later, Thompson and Wadsworth[48] published their classic paper describing the same process, admittedly in more detail, but essentially doing no more than confirming the findings of Ewing and Humphrey. Thompson and Wadsworth observed the formation, broadening and eventual development of surface cracks in slip bands on copper and nickel. This is today accepted as the sequence of events (Fig. 2.1) leading to the development of a surface crack in a ductile metal[29] when it is tested at room temperature at a stress level producing failure after a reasonable number of cycles, say, 10^5 or more. Because slip lines traverse a grain, cracking came to be termed transgranular.

In the 1950s, the concept of a two-stage process in the growth of a fatigue crack emerged. Mainly as a result of the metallographic work of Forsyth,[49] crack growth was divided into a Stage I crack and a Stage II crack. A Stage I crack was a slip band crack growing on planes of high shear stresses. It became a Stage II crack when it reached some length, changed direction and grew normal to the maximum principal tensile stress. Its growth rate was now much faster than that of a Stage I crack. More descriptive terms for these two stages, microcrack and macrocrack respectively, are sometimes used. Unqualified references to fatigue crack growth in this book and (usually) elsewhere are to macrocrack growth.

At about the same time, metallurgists were actively developing fractography, which is the study of fracture surfaces by optical[50] and electron microscopy[51] techniques. In the case of fracture surfaces created by a growing macrocrack, line markings, termed striations,

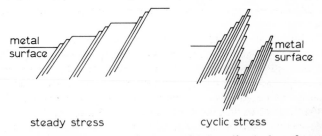

steady stress   cyclic stress

Fig. 2.1 Formation of surface cracks by slip (taken from Forsyth[29])

were seen on the surface, which indicated the position of the crack tip on successive load cycles (Fig. 2.2). Comparison of the distance between successive striations with measured crack growth rates showed that a crack tip moved forward by an increment of growth (the distance between adjacent striations) in each stress cycle. In retrospect, it is surprising that striations were not discovered earlier; the photograph reproduced as Fig. 2.2 was taken on a metallurgical microscope made in the 1930s. On a macroscopic scale, beach markings indicating changes in load can often be seen on fatigue fracture surfaces (Figs. 1.1 and 2.3); these are sometimes referred to, erroneously, as striations.

Metallographic work by Laird and Smith[52] elucidated the mechanism of striation formation. Essentially, plastic deformation at the crack tip changes the geometry of the crack tip during each load cycle. The nature of the mechanism of fatigue crack growth means that theories cannot be basesd on the accumulation of 'damage' ahead of the crack tip. The alternate blunting and resharpening of the crack tip explains why cracks can grow under cyclic loads too low to cause failure under a single load application. Striations are not always observed on fatigue fracture surfaces,[38] but cracks still grow by the same basic mechanism except that growth is not necessarily continuous along the whole crack front.

A major effort in the 1920–50 period was concerned with the prediction of the effect of a notch on the fatigue limit.[46] Elastic analysis of notch profiles, such as that provided by Inglis[53] and Neuber,[54] formed the basis for comparing theoretical prediction and experimental data, although as a consequence of Neuber's analysis, a large amount of data was obtained from specimens containing a single-edged V-notch, a profile which is seldom encountered in practice. In general, it was found that the fatigue limit of a notched specimen was not always equal to the plain fatigue limit of the material divided by the elastic stress concentration factor for the notch.

One intriguing fact that finally emerged in the 1950s was the existence of non-propagating cracks.[55] Sharply notched specimens unbroken after testing for long endurances were found to contain small cracks which had not grown owing to the phenomenon now known as the fatigue crack growth threshold. Their existence heralded the start of an intensive period of study into the behaviour of fatigue cracks, and at last provided a satisfactory explanation of notched fatigue strength.[56] The use of stress intensity factors (Section 4.3) to analyse fatigue crack growth data provided a further boost.[57]

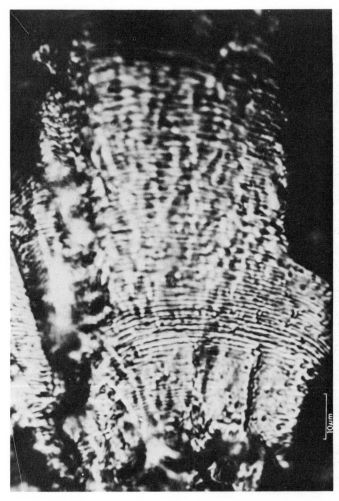

Fig. 2.2   Striations on the fracture face of an aluminium alloy specimen (constant amplitude loading; direction of crack growth left to right)

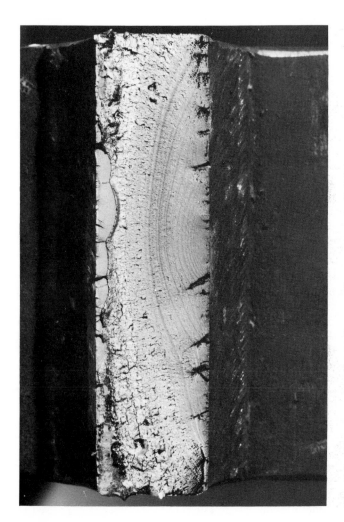

Fig. 2.3 Fracture surface of a medium strength C–Mn structural steel cruciform welded joint subjected to non-stationary narrow-band random loading

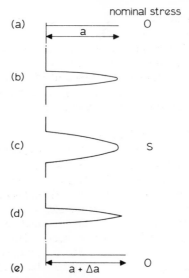

Fig. 2.4   Repeated sequence of crack opening and closing under cyclic loading, 0–S; the sequence (*a*) to (*e*) is repeated for each successive cycle

In retrospect, it is astonishing that very little experimental and no theoretical work was done on macrocrack growth until the 1950s. Griffith[58] published his classic paper on crack instability under a static load in 1920. Some 30 years later, in 1953, Head[59] published the first theory of crack growth under cyclic loading, only to find that there were no reliable crack growth data available which he could use to check his theory.

Using a simple model based on the sequence shown in Fig. 2.4, Frost and Dixon[60] developed a crack growth theory. This emphasized that resistance to crack growth was largely controlled by Young's modulus, rather than by a material's strength. At this time, it became clear that a threshold for fatigue crack growth existed because it was not physically possible for a crack to grow by less than about one lattice spacing per cycle. However, because crack growth is not necessarily continuous along the whole crack front, lower average rates are sometimes observed. The theory was later re-expressed in stress intensity factor form[61] (Section 4.9). As a geometry change is involved, any theory of crack growth must use an incremental rather than a deformation theory of plasticity,[62] and general features of the results can be predicted using dimensional analysis.[63] Because of the experimental difficulties involved in its study, the precise mechanism of microcrack growth is still unclear.

Modern work on the mechanisms of crack initiation and growth is concerned largely with the examination of detail, which can be complex, with a view to the development of alloys with improved resistance to fatigue.[64] The proceedings of a recent symposium[65] provide a convenient survey of current views.

2.4 CUMULATIVE DAMAGE

For many years, the vast majority of fatigue tests were carried out under constant amplitude loading, and designers were faced with the problem of how to use the resulting data to predict fatigue lives under the wide range of variable amplitude load histories encountered in service. The investigation of fatigue under varying stress amplitudes came to be known as the study of cumulative damage because of early interest in how fatigue 'damage' accumulated at various stress levels. Many attempts have been made to predict the life of a specimen in a cumulative damage test, and many data have been accumulated for the purpose of either deriving appropriate empirical relationships or testing theoretical predictions. Although the fundamental mechanisms are the same as under constant amplitude loading, in detail the problem is extremely complicated, even if only the crack propagation phase is considered (Section 6.5). In general, only empirical solutions to particular problems are possible.[66]

In 1924, Palmgren,[67] in predicting the life of ball-bearings, assumed that damage accumulated linearly with the number of revolutions. Similarly, in 1945 Miner[68] suggested that, in a fatigue test, damage to a specimen at a given stress level could be considered to accumulate linearly with the number of stress cycles, with failure occurring when the accumulated damage reached some critical value. Expressed mathematically, failure occurs when:

$$\sum \frac{n_i}{N_i} = 1 \qquad (2.1)$$

where n_i is the number of cycles at the ith load level, and N_i is the number of cycles to cause failure at the ith load level, taken from a constant amplitude S/N curve.

In this form, equation (2.1) is referred to as the linear cumulative damage law, the Palmgren–Miner law, or simply as Miner's rule. In practice, as recognized by Miner, the summation can differ widely from 1. Numerous modifications to Miner's rule have been proposed in attempts to obtain more accurate predictions, but none is of general applicability.

In 1956, Corten and Dolan[69] introduced the concept of a hypothetical (sometimes called effective) S/N curve, in which the constant amplitude S/N curve is adjusted so that the Miner's rule summation equals 1. The precise form of the hypothetical S/N curve is strongly dependent on the load history involved. The simplicity of Miner's rule makes it very convenient for design purposes, and it is widely used; the S/N curves quoted in standards etc. for use with Miner's rule, whatever the claimed derivation, should be regarded as hypothetical S/N curves, which may be further adjusted by appropriate safety factors to ensure that a conservative result is obtained.

Where the load history is such that individual load cycles cannot be distinguished, the usual approach is to use a cycle-counting rule to convert the load history to equivalent cycles.[70] The best known cycle-counting rule is the rainflow method (Section 6.5), which is now being incorporated in various standards such as BS 5400.[27]

2.5   FATIGUE TESTING AND DATA COLLECTION

Fatigue testing of specimens and structures goes back well over a hundred years. Repeated loading tests were carried out on chains and beams around the middle of the last century; Sir William Fairbairn published the results of his classic repeated bending wrought-iron beam tests in 1864.[71] All such tests showed that failure of the part occurred at a maximum load in the repeated loading cycle less than that required for static failure. The fatigue testing of specimens, and codification of the resulting data, can be said to have started with Wöhler who, around the 1850s, carried out his now classic experiments,[72] which led to the determination of S/N curves. On the continent these are often still called Wöhler curves. Wöhler controlled load accurately, found iron had a definite fatigue limit (i.e. unless a certain stress was exceeded, a specimen would not break no matter how many times this stress was applied) and studied the effect of mean stress. Bauschinger[73] extended Wöhler's work to other materials; Gerber[74] deduced his parabolic mean stress relationship from both Wöhler's and Bauschinger's data, as did Goodman[75] his straight-line relationship. Virtually all the data were concerned with tensile mean stresses, but both Gerber and Goodman, knowing that the static yield stress in compression equalled that in tension, assumed that a compressive mean stress had the same effect on the fatigue limit as a tensile mean stress. Hence, their relationships are symmetrical about zero mean stress. It is now well documented[15] that a compressive mean stress causes an increase in fatigue limit; nevertheless, the Goodman and Gerber relationships are still used today to predict the effect of a tensile mean stress; actual data usually fall between the two predictions. By the turn of the century, tests to determine the fatigue life of laboratory-type specimens were well established and a great deal of data had been organized into convenient forms.

Testing of laboratory specimens continued through the 1920s and '30s and considerable empirical data were accumulated on factors such as the effect of surface finish and specimen size. The popularity of specimen tests arose from two facts.[46] Firstly, the materials tested were ductile metals; these permitted the design and manufacture of cheap specimens which did not fail in the machine grips. Secondly, simple rotating bending fatigue-testing machines (Fig. 2.5) were readily available; these were cheap, accurate and reliable. It was certainly easier and cheaper to make a specimen than to make grips to hold a component or structural member in a standard test machine of that period. Occasional papers on fatigue crack growth rate measurements were published, cf. Ref. 76, but data were sparse and presented unanalysed. There was also interest in bulk specimen measurements, such as the measurement of hysteresis loops and internal damping with a view to developing so-called fundamental theories of fatigue.[8,9] For example, it was widely held that a material could absorb only a certain amount of energy before fracture occurred.

A major effort in the 1925–55 period[11,12] centred around the testing of notched specimens; the purpose was to predict the fatigue limit of a component having a change in section. Ample opportunity to hypothesize was provided when it was found that the fatigue limit of

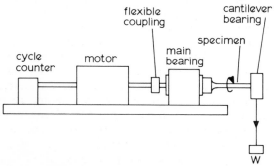

Fig. 2.5   Schematic of a rotating bending fatigue machine (from Gurney[34])

a notched specimen was not always equal to the plain fatigue limit
of the material divided by the elastic stress concentration factor for
the notch. Many laboratory fatigue data purported to show that
notched specimens of some materials could withstand, without breaking, nominal cyclic stresses giving rise to local stress amplitudes at
the notch root greater than the plain fatigue limit, whereas similarly
notched specimens of other materials could not. This led to the
former materials being termed notch insensitive and the latter notch
sensitive; the notch sensitivity index q given by

$$q = \frac{K_f - 1}{K_t - 1} \tag{2.2}$$

was often used to quantify this sensitivity, where K_f is the fatigue
strength reduction factor (fatigue limit for plain specimens/fatigue
limit for notched specimens based on nominal stress) and K_t the
elastic stress concentration factor (actual/nominal stress) for the
notch root. However, q is not a material constant[46] and its use is to
be discouraged as it can lead to unsafe predictions. Another favoured
topic[11,12] was combined stresses, especially attempting to predict the
ratio of the torsional to uniaxial fatigue limits. Satisfactory explanations of both notched specimen and combined stress experiments
had to await understanding of the role of fatigue crack growth.[46]

By the end of this period,[14] there was increasing interest in structural fatigue testing with increasingly elaborate equipment, and in
'low cycle' fatigue of specimens where the total life was less than
about 10^5 cycles. Because of plasticity effects,[32] low cycle fatigue tests
are usually carried out under constant strain rather than constant
load conditions, with the results plotted as strain/life curves. Sometimes, the whole S/N curve is plotted in terms of strain rather than
stress to present a unified picture.[7]

In the modern era of fatigue testing, taken as the mid-1950s to
the present day, three main trends are discernible: structural fatigue
testing, tests to determine fatigue crack growth behaviour, and the
codification of large amounts of experimental data. The art of fatigue
testing has now reached the stage where a standard text on good
practice is available,[35] and national standards on the fatigue testing
of materials have been developed in industrialized countries.[77,78]

The Comet accidents in 1954[79] stimulated extensive fatigue testing
of aircraft structures. Structural fatigue testing using elaborate equipment is now routine in the aerospace industry.[35,80,81] In the 1950s,
the equipment available for structural fatigue testing was relatively
crude. The situation changed with the development of servohy-

draulic-controlled actuators,[35] which permitted the application of virtually any desired load history such as broad-band random loading (Section 6.5). The development of cheap microprocessors[82,83] and strong floor testing facilities[84] has stimulated structural fatigue testing in general engineering. Figure 2.6 shows a recent test on a crane boom using a strong floor facility.[84] Lifting and slewing loads were simulated by two long-stroke hydraulic actuators, mounted at 45°, in a microprocessor-controlled sequence. The feasibility of such tests is nowadays controlled by financial rather than technical constraints. Despite their importance,[66,84] procedures for structural fatigue testing have not yet in general been formally standardized, although the occasional *ad hoc* test is specified for load-bearing structures where trouble has been experienced. For example, BS 5153[85] includes a fatigue test for the heels of ladies' shoes. Some progress has been made in the development of standardized load histories for the testing of important structures.[86,87]

Over this modern period, the increasing availability of machines capable of applying variable amplitude loads ranging from simple two-level tests to elaborate simulation of service histories has resulted in large numbers of tests being carried out on relatively simple specimens. Usually, the aim is to investigate some aspect of cumulative damage (cf. previous section), but the wide range of possible load histories makes it difficult to extract useful generalizations.[66]

Fatigue crack growth tests are relatively straightforward. Consequently, following the appreciation of their importance, large numbers of such tests have been carried out, mostly under constant amplitude loading. Reference 88 is a bibliography of data for a wide range of metals and alloys. At first sight, data reduction is also straightforward, but it does present some subtle and intractable problems.[89] Because of this, a standard method of test[90] has only recently been developed. Measurement of the threshold for fatigue crack growth is again fairly straightforward, but the actual value obtained is quite strongly dependent on the precise method used,[38] and no standard has so far appeared. Large numbers of simple, variable amplitude, fatigue crack growth rate tests, for example the injection of an occasional high load into a constant amplitude sequence, have been carried out in optimistic attempts to acquire basic information which could be used to predict behaviour under more complicated load histories. There is increasing interest[38] in the fatigue behaviour of cracks in biaxial stress fields, and cracks inclined to the applied stress. This is analogous to the earlier interest in the combined stress behaviour of uncracked specimens.

Fig. 2.6   Fatigue test of a crane boom

In any largely descriptive subject, such as metal fatigue, there is a need to organize observations so that they are more intelligible and useful. Experimental fatigue data given in research papers are rarely in a form which can be applied directly to practical engineering problems, and considerable effort has been devoted to the analysis and codification of such data. For example, the Engineering Sciences Data Unit has issued a number of data sheets on metal fatigue.[91] Some of these are based on extensive specially commissioned tests such as those on bolted joints.[92] The new British Standard for the fatigue design of bridges[27] was based on extensive fatigue test data on welded joints.[33]

An unfortunate trend in recent years has been the increasing number of papers describing work that has no clear objective. Some of these tests can neither be described as research in the sense that they are likely to lead to deeper insight into the nature of metal fatigue, nor as data collection in that the information is required for a specific purpose. Usually the work described is only new in that some particular combination, often obscure, of material, specimen design and test has not been tried before. Frequently, so-called theory turns out to be little more than a curve-fitting exercise, resulting in equations, which, like equation (2.2), have only limited applicability. Fatigue testing is time consuming and expensive, but papers, such as Ref. 93, which claim to provide short-cuts to the acquisition of fatigue data, should be treated with caution as they are usually based either on empirical correlations of limited applicability or an over-simplified view of the nature of fatigue.

REFERENCES

1   J. Y. MANN: 'Bibliography on the fatigue of materials, components and structures, Vol. I, 1838–1950', 1971; 'Vol. II, 1951–1960', 1978; Oxford, Pergamon Press

2   S. P. TIMOSHENKO and J. N. GOODIER: 'Theory of elasticity', 3rd ed., 1970, New York, McGraw-Hill Book Company

3   R. J. ROARK: 'Formulas for stress and strain', 4th ed., 1965, New York, McGraw-Hill Book Company

4   A. H. COTTRELL: 'An introduction to metallurgy', 2nd ed., 1975, London, Edward Arnold (Publishers) Ltd

5   J. NEWTON: 'Introduction to metallurgy', 2nd ed., 1947, New York, John Wiley & Sons Inc.

6   C. BATHIAS and J.-P. BAÏLON (eds.): 'Le fatigue des matériaux et des structures' ('Fatigue of materials and structures'), 1980, Paris, Maloine SA

7   H. O. FUCHS and R. I. STEPHENS 'Metal fatigue in engineering', 1980, New York, Wiley–interscience

8   H. J. GOUGH: 'The fatigue of metals', 1924, London, Scott, Greenwood & Son

9   H. F. MOORE and J. B. KOMMERS: 'The fatigue of metals', 1927, New York, McGraw-Hill Publishing Co.

10 P. G. FORREST: 'Fatigue of metals', 1962, Oxford, Pergamon Press

11 H. J. GROVER, S. A. GORDON and L. R. JACKSON: 'Fatigue of metals and structures', 1954, Washington, DC, Department of the Navy

12 R. CAZAUD: 'Fatigue of metals' (trans. A. J. Fenner), 1953 London, Chapman & Hall Ltd

13 J. A. POPE (ed.): 'Metal fatigue', 1959, London, Chapman & Hall Ltd

14 Proc. Int. Conf. on Fatigue of Metals, 1956, London and New York; 1958, London, Institution of Mechanical Engineers

15 N. E. FROST, K. J. MARSH and L. P. POOK: 'Metal fatigue', 1974, Oxford, Clarendon Press

16 S. KOCÁNDA: 'Fatigue failure of metals', 1978, Alphen aan der Rijn, Sijthoff and Noordhoff International Publishers

17 M. KLESNIL and P. LUKÁŠ: 'Fatigue of metallic materials', 1980, Amsterdam, Elsevier Scientific Publishing Co.

18 A. H. COTTRELL: 'Theory of crystal dislocations', 1962, London, Blackie & Son Ltd

19 W. ALEXANDER and A. STREET: 'Metals in the service of man', rev. ed., 1951, Harmondsworth, Penguin Books Ltd

20 W. ALEXANDER and A. STREET: 'Metals in the service of man', 6th ed., 1976, Harmondsworth, Penguin Books Ltd

21 J. E. GORDON: 'The new science of strong materials', 1968, Harmondsworth, Penguin Books Ltd

22 W. C. UNWIN: 'The elements of machine design', 1894, London, Longmans, Green & Co.

23 K. J. PASCOE: 'An introduction to the properties of engineering materials', 3rd ed., 1978, London, Van Nostrand Reinhold Company

24 T. V. DUGGAN and J. BYRNE: 'Fatigue as a design criterion', 1977, London, Macmillan Press Ltd

25 C. C. OSGOOD: 'Fatigue design', 2nd ed., 1982, Oxford, Pergamon Press

26 W. J. HARRIS and G. SYERS: 'Fatigue alleviation', Engineering design guides No. 32, 1979, London, Oxford University Press

27 BRITISH STANDARDS INSTITUTION: 'Steel, concrete and composite bridges. Part 10: Code of practice for fatigue', BS 5400: Part 10: 1980

28 A. G. PUGSLEY: 'The safety of structures', 1966, London, Edward Arnold

29 P. J. E. FORSYTH: 'The physical basis of metal fatigue', 1969, London, Blackie and Son Ltd

30 L. G. JOHNSON:'The statistical treatment of fatigue experiments', 1964, London, Elsevier

31 W. J. HARRIS: 'Metallic fatigue with particular reference to the significance of certain standard aircraft fabrication and finishing processes', 1961, Oxford, Pergamon Press

32 B. I. SANDOR: 'Fundamentals of cyclic stress and strain', 1972, Madison, Wisc., University of Wisconsin Press

33 T. R. GURNEY: 'Fatigue of welded structures', 1968, London, Cambridge University Press

34 T. R. GURNEY: 'Fatigue of welded structures', 2nd ed., 1979, London, Cambridge University Press

35 S. R. SWANSON (ed.): 'Handbook of fatigue testing', ASTM STP 566, 1974, Philadelphia, Pa., American Society for Testing and Materials

36 R. E. LITTLE and E. H. JEBE: 'Statistical design of fatigue experiments', 1975, London, Applied Science Publishers Ltd

37 D. BROEK: 'Elementary engineering fracture mechanics', 1974, Leyden, Noordhoff International Publishing

38 G. G. CHELL (ed.): 'Developments in fracture mechanics—I', 1979, London, Applied Science Publishers Ltd

39 'A general introduction to fracture mechanics', 1979, London, Mechanical Engineering Publications Ltd

40 G. P. CHERAPANOV: 'Mechanics of brittle fracture' (in Russian), 1974, Moscow, Nauka Publishers; English translation with supplementary (1977) material, 1979, New York, McGraw-Hill

41 S. T. ROLFE and J. M. BARSOM: 'Fracture and fatigue control in structures. Applications of fracture mechanics', 1977, Englewood Cliffs, NJ, Prentice-Hall Inc.

42 N. SHUTE: 'No highway', 1948, London, William Heineman Ltd

43 G. PAWLE: 'The secret war 1939–45', 1956, London, George C. Harrap & Co. Ltd

44 R. H. PARSONS: 'History of the Institution of Mechanical Engineers 1847–1947', 1947, London, Institution of Mechanical Engineers

45 ANON: *Technical Report. British Engine Insurance Ltd*, 1978, **13**, 9–47

46 N. E. FROST: *J. Soc. Environ. Eng.*, 1975, **14-2**, (65), 21–24, 27–28

47 J. A. EWING and J. C. W. HUMPHREY: *Philos. Trans.*, 1903, **200**, 241–250

48 N. THOMPSON and N. J. WADSWORTH: *Adv. Phys. (Philos. Mag. Suppl.)*, 1958, **7**, (25), 62–169

49 P. J. E. FORSYTH: Proc. Crack Propagation Symp., 1961, Vol. I; 76–94, 1962, Cranfield, College of Aeronautics
50 C. A. ZAPPFE and C. O. WORDEN: *Trans. ASM*, 1961, **43**, 958–969
51 'Electron fractography', STP 436, Philadelphia, Pa., 1968, American Society for Testing and Materials
52 C. LAIRD and G. C. SMITH: *Philos. Mag.*, 1962, **7**, (77), 847–857
53 C. E. INGLIS: *Trans. Inst. Nav. Archit.*, 1913, **55**, 219
54 H. NEUBER: 'Theory of notch stresses', 1946, Ann Arbor, Mich., J. Edwards
55 N. E. FROST and C. E. PHILLIPS: as Ref. 14, pp. 520–526
56 N. E. FROST: as Ref. 14, p. 745
57 P. C. PARIS and F. ERDOGAN: *Trans. ASME, J. Bas. Eng.*, 1963, **85** , (4), 528–553
58 A. A. GRIFFITH: *Philos. Trans.*, 1921, **221**, 163–198
59 A. K. HEAD: *Philos. Mag.*, 1953, **44**, (7), 925–938
60 N. E. FROST and J. R. DIXON: *Int. J. Fract. Mech.*, 1967, **3**, (4), 301–316
61 L. P. POOK and N. E. FROST: *Int. J. Fract.*, 1973, **9**, (1), 53–61
62 B. BUDIANSKY: *J. Appl. Mech.*, 1959, **26**, (2), 259–264
63 J. R RICE: 'Fatigue crack propagation', STP 415, 247–309, 1967, Philadelphia, Pa., American Society for Testing and Materials
64 'Achievement of high fatigue resistance in metals and alloys', STP 467, 1970, Philadelphia, Pa., American Society for Testing and Materials
65 J. T. FONG (ed.): 'Fatigue mechanisms', STP 675, 1979, Philadelphia, Pa., American Society for Testing and Materials
66 W. SCHÜTZ: *Eng. Fract. Mech.*, 1979, **11**, (2), 405–421
67 A. PALMGREN: *VDI Z.*, 1924, **68**, (14), 339–341
68 M. A. MINER: *J. Appl. Mech.*, 1945, **12**, (3), A159–A164
69 H. T. CORTEN and T. J. DOLAN: as Ref. 14, pp. 235–246
70 P. WATSON and B. J. DABELL: *J. Soc. Environ. Eng.*, 1976, **15-3**, (20), 3–9
71 W. FAIRBAIRN: *Philos. Trans.*, 1864, **154**, 311–325
72 A. WÖHLER: *Engineering*, 1871, **11**, 199–200, 221, 244–245, 261, 299–300, 326–327, 349–350, 397, 439, 441
73 J. BAUSCHINGER: *Mitt. Mech.-Tech. Lab. Münch.*, 1886, **13**, 1
74 W. GERBER: *Z. Bayer. Archit. Ing. Ver.*, 1974, **6**, 101
75 J. GOODMAN: 'Mechanics applied to engineering', 1899, London, Longmans, Green & Co.
76 A. V. DeFOREST: *J. Appl. Mech.*, 1936, **3**, A23–A25
77 BRITISH STANDARDS INSTITUTION: 'Yearbook 1980', 1979, London, British Standards Institution

78   1980 annual book of ASTM standards, Part 10', Philadelphia, Pa., 1980, American Society for Testing and Materials

79   'Report on Comet accident investigations. Accident Note No. 263. Part 3: Fatigue tests on the pressure cabin and wings', 1954, Farnborough, Royal Aircraft Establishment

80   W. T. KIRKBY, P. J. E. FORSYTH and R. D. J. MAXWELL: *Aeronaut. J.*, 1980, **84**, (829), 1–12

81   W. G. HEATH: *ibid.*, 1980, **84**, (31), 81–92

82   F. SHERRATT and P. R. EDWARDS: *J. Soc. Environ. Eng.*, 1974, **13-4**, (63), 3–14

83   W. LUCAS and K. SLOAN: *Weld. Inst. Res. Bull.*, 1978, **19**, (11), 331–339

84   K. J. MARSH: *Int. J. Fatigue*, 1979, **1**, (1), 3–6

85   BRITISH STANDARDS INSTITUTION: 'Methods of test for footwear and footwear materials. Section 4.9: Fatigue resistance of heels of ladies' shoes', BS 5131 : Section 4.9 : 1975

86   'FALSTAFF (Description of a Fighter Aircraft Loading STAndard For Fatigue evaluation)', 1976, Darmstadt, Laboratorium für Betriebs-festigkeit

87   L. P. POOK: *J. Soc. Environ. Eng.*, 1978, **17-1**, (76), 22–23, 25–28, 31–35

88   C. M. HUDSON and S. K. SEWARD: *Int. J. Fract.*, 1982, **20**, (3), R59–R117

89   L. P. POOK: *J. Soc. Environ. Eng.*, 1976, **15-4**, (71), 3–10

90   ASTM E647–8: 'Standard test method for constant-load-amplitude fatigue crack growth rates above 10^{-8} m/cycle', 1981, Philadelphia, Pa., American Society for Testing and Materials

91   'Index 1977–78', 1978, London, Engineering Sciences Data Unit

92   R. H. SANDIFER: 'Bolted joint research programme. S & T Memo 1-77. DRIC-BR-60447. Vols. I–IV', 1977–78, St Mary Cray, Kent, Defence Research Information Centre

93   P. R. WEIHSMANN: *Mater. Eng.*, 1980, **91**, (3), 52–54

CHAPTER 3

The basis of design against metal fatigue

It was not time to shut the stable door
when the horses be lost and gone.
William Caxton, *Fables of Aesop*.

3.1 INTRODUCTION

It is easy for fatigue specialists to forget that fatigue is only one of
the many factors that need consideration during the design of an
engineering structure. It is probably fair to say that fatigue behaviour
is usually checked only after the basic design has been decided, and
there must be many successful structures which were designed with
no specific attention having been paid to fatigue. In one sense there
is no such thing as fatigue design, only fatigue assessment of existing
designs. Often, a fatigue assessment will reveal the need to modify
the original design. In structures where the possibility of fatigue is
a major preoccupation of the designer, the two are integrated.

Provision for the attachment of accessories to load-bearing parts,
made after the fatigue assessment has been carried out, is a frequent
cause of fatigue failure. For example, the capsize of the semi-submers-
ible accommodation rig *Alexander L. Kielland* was attributed[1,2] to
fatigue at the attachment for a hydrophone used to position the rig
on site. The hydrophone was installed on the underside of a 2·4 m
diameter horizontal bracing member (Fig. 3.1). To install the hydrophone, a 290 mm diameter steel tube was welded into a hole cut into
the 25 mm thick brace so that it protruded by about 300 mm. The
hydrophone diaphragm assembly was bolted to a 100 mm wide flange
at the end of the tube. A fatigue crack had spread from the weld
attaching the tube around about one-third of the brace's circumfer-
ence before ductile tearing completed the failure[1,3] (Fig. 3.2).

Even rough estimation of fatigue strength at the preliminary design
stage presents problems. Why should this be, since large amounts of
experimental data are available and, in general, the basic mechanisms
of fatigue failure are relatively well understood and documented?

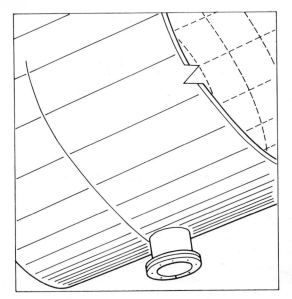

**Fig. 3.1   Mounting of the hydrophone on a cross-bracing
member of the *Alexander L. Kielland* (from
Cottrill[2])**

Indeed, in broad terms, if not in detail, the mechanisms are not very
complex. Problems arise because of the complexity and variety of
realistic engineering design situations, and the difficulties of gen-
eralizing on known fatigue data and then applying these generalizations to a specific engineering problem.[4] There is in general no
simple relationship between a metal's fatigue resistance and its tensile
strength, so simple substitution of a stronger material is seldom the
answer to a fatigue problem. At times there appears to be little
correlation between what appear to be flaws of one sort or another
and service performance.[5]

Most books on fatigue contain advice on its alleviation (Section
2.2). Much of this amounts to hints and tips on how to reduce the
severity of fatigue stresses at actual or potential failure sites. Many
fatigue problems are solved by taking some action to improve fatigue
performance on what amounts to a trial-and-error basis,[6] but knowledge and experience are required. It is fatally easy to provide a 'cure'
which is worse than the disease, especially when welding is involved,[7]
and the results of failure analyses can leave designers with difficult

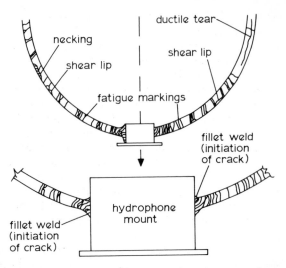

Fig. 3.2   Fracture surface of a cross-bracing member of the *Alexander L. Kielland* (from *Offshore Eng.*[3])

decisions (Section 3.2.1). A knowledge of whether cracks are present is of major importance in predicting behaviour under fatigue loading. Indeed, from a practical viewpoint, probably the most significant advance in metal fatigue during the past few years[4] is the general realization that many structures contain crack-like flaws which are either introduced during manufacture, especially if welding is used,[7] or form early on during service.[8] Fatigue crack growth from these flaws may occur during virtually the whole life of the structure. An understanding of fatigue crack growth is therefore essential for the understanding and prediction of the fatigue behaviour of such structures, and is emphasized in the remainder of this book.

3.2   FAILURE ANALYSIS

Many authors (see, for example, Refs. 6 and 9–12) have drawn attention to the importance of analysing fatigue failures so that future designs may be modified to avoid similar failures. The Proceedings of the Institution of Mechanical Engineers' Conference on Fatigue, held in 1956,[12] include about 100 examples of fatigue failures, all intended to illustrate some point on the alleviation of fatigue. Precise classification of the reason for a failure is a matter of semantics. Final failure takes place when a fatigue crack has grown so far that the section is so reduced that a simple overload failure occurs or the conditions for a brittle fracture (Section 4.3) are satisfied. Relatively small fatigue cracks can be the source of extensive low stress brittle fractures in structural steels.[13] Such failures are usually described as being due to brittle fracture. Many fatigue failures, perhaps the majority, are never diagnosed. Failures such as those shown in Figs. 1.1–1.4 are only recognized as being due to fatigue if they happen to be seen by a specialist. Unrecognized fatigue failures are probably the underlying cause of some apparently inexplicable accidents, especially those involving motor vehicles for which drivers could well be wrongly blamed. It is at times difficult to establish the precise sequence of events leading up to a particular failure. The reasons range from simple forgetfulness, through lack of adequate documentation, to a human desire to avoid blame.

Using the word in the legal (Section 3.5) rather than intuitive sense, defects leading to fatigue failure may be divided into two broad classes. Firstly, 'design defects' where the original design was inadequate for the envisaged use, including cases where the loading was more severe than anticipated by the designer; and secondly, 'manufacturing defects' where the failure occurred because the designer's intentions were not followed during manufacture, including faulty repairs and the use of material that is incorrect or below specification. Relatively few fatigue failures are found to be due to the use of faulty or incorrect material.

One of the earliest documented examples of failure analysis is in the records of the Institution of Mechanical Engineers.[14] Over 100 years ago, members were concerned with the fracture of wrought-iron railway axles. One member had the acumen to notice that all axles failed (by fatigue) at the shoulder on the axle used to locate the wheels, and rightly stated that it was more important to redesign the axle to eliminate this shoulder and so reduce stresses, than to argue about the mechanisms of fatigue. From a practical design viewpoint this advice still holds good.

3.2.1 Failure analysis of a large cast-steel shaft

Figure 3.3 shows a section through a large cast-steel shaft which failed after $2\frac{1}{2}$ years' service. The steel used had a tensile strength of $850\,\mathrm{MN\,m^{-2}}$ and, as is usual with large steel castings,[15] some repair welding was carried out to rectify inevitable casting defects. The equipment was intended for continuous operation, with downtime kept to a minimum to avoid heavy consequential costs. The shaft failed at the 15 mm radius shown in Fig. 3.3. Examination showed that the repair welding had not been completely successful and fatigue cracks had grown from several crack-like flaws roughly 1 cm deep. These had joined up and extended most of the way through the section before final failure took place.

Loading on the shaft was essentially rotating bending due to self-weight. When some additional loads and the stress concentration factor were taken into account, the stress at the failure site was several thousand cycles/day at $3 \pm 108 \text{ MN m}^{-2}$. The designer assumed, wrongly, that he was dealing with an uncracked component and, noting that this and stresses at other potentially critical locations were well below the plain specimen fatigue limit of the material, concluded that the shaft would have an indefinite fatigue life.

However, a simple fracture mechanics calculation (Section 4.7) showed that the maximum permissible flaw depth for indefinite life was 0·37 mm, which is not much larger than the unavoidable crack-like flaws associated with welding.[7] The much deeper flaws actually present made eventual failure inevitable. From the operator's viewpoint, a replacement shaft with indefinite life was obviously preferable, but this is virtually impossible to achieve. Flaws of the calculated critical size are too small to be detected in cast steel by non-destructive inspection.[16] Feasible redesign would not produce significant reduction in stresses at critical locations. As cracks are bound to be present and crack growth properties are largely independent of a steel's tensile strength (Section 4.5), substitution of a stronger steel would not produce any improvement. Replacement with a forging would reduce the probability of unacceptable flaws being present, but would be expensive, and detection of critical flaws would still not be feasible.

From the manufacturer's viewpoint, the simplest solution would be to supply an identical shaft and persuade the operator to accept that it had a limited life. It could either be replaced after a limited amount of use, or inspected regularly and replaced if cracks had developed to more than a certain depth. In either case, a fatigue assessment (see next section) would have to be carried out to determine appropriate replacement or inspection intervals to ensure that unexpected failure did not occur. Unfortunately, either replacement or inspection would involve a considerable amount of dismantling, with concomitant expensive downtime, so it is clear that there is no easy solution to this particular fatigue problem. However, failure of the shaft is not actually dangerous, so possibly the best solution in this particular case would be simply to keep a replacement on hand and replace the shaft when it fails.

3.3 FATIGUE ASSESSMENT:
SITUATIONS, APPROACHES AND PHILOSOPHIES

In approaching a mechanical engineering design problem in general, or a fatigue assessment in particular, a designer must first select, perhaps unconsciously, the approach to be used. Failure to adopt an appropriate overall strategy, including allowance for human fallibility, is the one theme common to diverse catastrophic failures.[17] The topic is frequently discussed in the aerospace literature, e.g. Ref. 18, and Ref. 6 gives some examples from general engineering. Arguments over the selection of an appropriate overall strategy are at the heart of the controversy over the safety of nuclear power stations.[19] At a more mundane level, much time and expense can often be saved by choosing an appropriate strategy early in the job. The planning of modifications and repairs should be taken as seriously as the original design. Too often it is left to skilled tradesmen who lack the necessary expertise.[20]

3.3.1 Situations

There are four basic situations in mechanical engineering design which will often include a fatigue assessment.[21] The first is the 'in-house tool'. A tool engineer given the job of designing a welding fixture for some new device will probably think about it, sketch several possible solutions, select one by intuition or with the help of some quick analysis, and proceed to draw the tool which will do the desired job. He relies on his ability to solve quickly any problem which arises when the tool is first used. He will do little or no stress analysis or optimization. In effect, he is relying on previous experience of broadly similar jobs and the assumption that any failure will not be catastrophic.

Second is the 'mass product'. The product engineer in charge of a new digital door lock will probably build several prototypes, test them extensively, and analyse and optimize the design as much as possible. The cost of design and analysis can be amortized over a large number of products, so it is worth evaluating several ideas thoroughly. The eventual solution will be documented in detail. Figures 1.1–1.4 are all examples of components from mass products.

Thirdly, there is the '£1 000 000 machine'. The designer of a 100 MW steam turbine cannot afford prototypes; he must learn as much as possible from existing field experience and make much use of analysis, even though it be complex and expensive. The cost of design and analysis will normally be small compared with rectification of faults at a later stage. The failure of the large cast shaft, described in Section 3.2.1, exemplifies the difficulties which can result from inadequate analysis at the design stage.

The fourth is the 'code design'. The designers of a boiler must pay more attention to codes than the other designers mentioned. If the code specifies a limiting stress and methods for calculating it, this designer, in contrast to the others, may need to compute stress to three significant figures. Also, he is less concerned with checking whether the written requirements are the real requirements, and more concerned with legalistic study of the code, which he assumes will lead to a satisfactory design. Section 3.4 gives an example of a fatigue assessment using codes. Most mechanical design jobs fall somewhere between these extremes. A motor car is a mass product, but subject to various codes.

3.3.2 Approaches

There are three main approaches to fatigue assessment: an analytic approach, use of a standardized procedure, and service-loading testing. In practice, some combination of these is normally used. Whatever approach is used, allowance must be made for the inevitable scatter in the fatigue life of specimens and structures (Chapter 5), and factors such as uncertainties in service load history and stress analysis.[9] Methods range from a 'safety factor' based on experience, through the statistical methods[7] used to develop the design curves given in BS 5400,[22] to the elaborate procedures used in the assessment of nuclear pressure vessels.[23] The degree of confidence required obviously depends on the consequences of a failure. For example, the failure of the rotating head of a centrifuge would be more serious if it were being used in a medical laboratory and dangerous organisms escaped.

An analytic approach makes use of information on service loads, material properties and applied mechanics. Such an approach requires expert knowledge. The crack initiation and propagation phases must be considered separately, although in practice one or the other normally predominates. This is usually the crack propagation phase; the exceptions are generally carefully made, highly stressed components where stress concentrations are kept to a minimum.[24] The failure analysis of the large cast shaft in Section 3.2.1 is a simple example of an analytic fatigue assessment. More typically, elaborate calculations are necessary, and the approach may fail because not all the large amount of detailed information required is available.[25] The desire to put fatigue assessment on an analytic basis has stimulated much of the academic work which has been carried out on fatigue; some is described in the next four chapters.

From the designer's viewpoint, standardized procedures of various degrees of formality are undoubtedly the most satisfactory. These range from informally established 'good practice' in a particular design office to elaborate published codes, often imposed by regulatory authorities. They may be based on analytic procedures, service experience, or on both, and need only have an empirical basis provided they give sufficiently accurate answers. For the large cast shaft described in Section 3.2.1, a standard permissible fatigue stress could have been derived by consideration of the maximum expected flaw size. This flaw size would be established on the basis of what could be achieved by good practice, combined with the sensitivity and reliability of available non-destructive test equipment. The use of more stringent quality control and non-destructive test procedures would reduce the maximum expected flaw size and hence increase the permissible fatigue stress. All codes are based on compromises of this type. Differences in detail design requirements, and in the stringency of quality control required, usually account for the different stresses permitted for similar structures by different codes.[26] Standardized methods have the advantage that little or no expert knowledge is required and they can be conveniently incorporated in software packages for computers. This facilitates the assessment of complex structures.[27] Available packages for fatigue design should be carefully examined to ensure that the basis on which they operate is suitable for the intended application.

Modern servohydraulic equipment permits the application of virtually any load history (Section 2.5) and can therefore be used to determine service life where analytic and standard methods fail. Service-loading testing of prototypes is being increasingly used for critical structures[28–30] in various branches of engineering, and is sometimes a requirement of regulatory authorities. It has the advantage that its basis is easily understood by laymen. Acquiring the necessary information on service loads may be a major problem, and expert judgement is needed in deciding how representative the loading applied to the structure needs to be. Actual load histories in service cover a very wide range of possibilities and can be difficult to characterize because they are usually statistically non-stationary (Section 5.2); Fig. 3.4 shows an example taken from railway practice.[31] Such histories usually require processing before they can be used for test purposes.[32] Various simplifications and compromises are usually necessary when load histories for test purposes are calculated theoretically; Fig. 3.5 shows a load history developed for testing

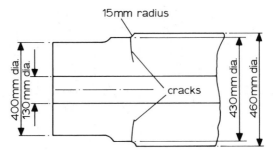

Fig. 3.3 Section through a large cast-steel shaft

relevant to offshore structures.[33] Each sub-block is a stationary narrow-band random loading (Section 5.2). In the aircraft industry, major efforts are devoted to the development of standardized load histories for test purposes.[25,34] Structural fatigue testing has the advantage that weak points in a design can be identified and rectified at the prototype stage. Simple comparative tests are sometimes useful in establishing whether a proposed modification does result in a worthwhile increase in fatigue strength.

The optimum approach to a particular problem is a judicious blend of theoretical calculations, experiment and analysis of relevant service experience. Often, analytical methods are best used to extrapolate experimental data to broadly similar situations. When innovation is involved, ensuring the integrity of structures subject to fatigue loading can be an extremely expensive, time-consuming

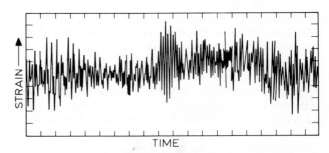

Fig. 3.4 Example of service strain history for a fabricated rail vehicle component (from Dabell and Watson[31])

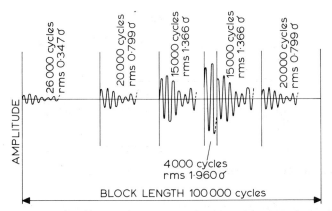

AMPLITUDE

26 000 cycles
rms 0·347 σ

20 000 cycles
rms 0·799 σ

15000 cycles
rms 1·366 σ

15000 cycles
rms 1·366 σ

20000 cycles
rms 0·799 σ

4000 cycles
rms 1·960 σ

BLOCK LENGTH 100 000 cycles

**Fig. 3.5 Load history for testing relevant to offshore
structures (from Pook[33])**

business. The United Kingdom Offshore Steels Research Project
(UKOSRP)[35] is an example. It was set up to collect fatigue and
fracture information relevant to tubular structures in the North Sea
where wave loading is very severe. Some results obtained so far[36]
suggest that existing design codes may need modification. The
UKOSRP is a major part of extensive European collaborative work[37]
on fatigue of offshore structures.

3.3.3 Philosophies

The two basic philosophies of fatigue design ('safe life' and 'fail-safe')
were originally developed in the aircraft industry.[38] In a structure
designed on 'safe life' principles, the object is to ensure that the
structure will not fail within the design life, which may be indefinite.
When the design life is reached, the part concerned is discarded.
The safe life may be extended by periodic inspection for cracks. If
one is found the part is discarded. In this case, a fatigue assessment
must have demonstrated that the largest crack likely to be missed
by the inspection technique used must not lead to failure before the
next inspection. Normally, steps to ensure that adequate inspection
is feasible need to be taken at the design stage. The large cast shaft
described in Section 3.2.1 is an example of a safe life design. In
retrospect, inspection of the shaft should have been allowed for at
the design stage, and inspection intervals decided from an assessment
of the rate of fatigue crack growth. If the part concerned is cheap,
such as the rotating head of a centrifuge, the cost of inspection may

well be greater than the cost of simply replacing the part. Again, it may be cheaper to express replacement intervals in time, based on the maximum likely rate of usage, rather than to go to the trouble of keeping detailed records.

In a 'fail-safe' structure, alternative load paths are provided, and the structure is redundant in that the failure of any one member does not lead to failure of the whole structure. Once the structure has lost redundancy owing to failure of a member it is less safe and provision must be made, perhaps by regular inspection, for this loss of redundancy to become obvious. In practice, and tubular offshore structures are an example,[27] many structures are designed using a combination of fail-safe and safe life philosophies. The incorporation of fail-safe features makes the fatigue assessment of safe life less crucial, but too much reliance should not be placed on fail-safe features. The failure of a cross-bracing member on the accommodation rig *Alexander L. Kielland* (Section 3.1) resulted in catastrophe because, in the event, the structure proved to be non-redundant.[1,3] Proposed modifications to sister rigs include additional cross-bracing members to ensure redundancy.[3]

A variation on fail-safe design is to ensure that failed parts are constrained before they can cause injury or excessive damage. For example, failure of the rotating head of a centrifuge, if its contents are innocuous, is merely an expensive nuisance, provided the broken bits are contained within the machine.

The 'damage-tolerant' approach to aircraft design has recently been developed[30,39] as a supplement to the safe life and fail-safe philosophies. It requires that fatigue assessments be carried out using the assumption that cracks of certain specific sizes are present at all critical locations such as rivet holes. It therefore directs particular attention to the determination of crack growth rates in structures.

3.4 EXAMPLE OF FATIGUE ASSESSMENT

Most fatigue assessments are based on simplified, standardized procedures which, despite their apparent lack of physical validity, are known to give conservative answers. An example of this type is therefore given in some detail.

Flow measurement facilities at the National Engineering Laboratory[40] include a unique high pressure gas flow primary standard test rig. The main air storage vessel for this rig is a 1·8 mm diameter welded steel cyclinder 6·3 m long which has a volume of 11·5 m^3, and it was intended to hold air at pressures up to 138 bar. The consequences of failure are therefore potentially catastrophic. The vessel was manufactured to BS 1515 : Part I : 1965,[41] which was written for Imperial units and these are therefore used below.

The estimated fatigue loading for the vessel in terms of internal pressure was:

0–2 000 lbf/in^2 (0–137·9 bar):
100 cycles/year (full pressure cycles)

1 400–2 000 lbf/in^2 (96·5–137·9 bar):
1 000 cycles/year (fluctuating loading)

which gave for the 15 year design life:

0–2 000 lbf/in^2: 1 500 cycles

1 400–2 000 lbf/in^2: 15 000 cycles

Perusal of the manufacturer's drawings and calculations during commissioning in 1976 showed that the vessel met the requirements of BS 1515, except that the full pressure cycles had not been taken into account, and the number of fluctuating cycles was taken as 4 cycles/day, i.e. 21 900 in the 15 year design life. (According to BS 1515 it is the purchaser's responsibility to ensure that full information on fatigue loadings is made available to the manufacturer.) Further calculations were therefore carried out to BS 1515, and also to the more recent BS 5500 : 1976.[42] In all cases, the maximum design temperature was taken as 50°C.

The manufacturer had carried out a simplified fatigue analysis in accordance with BS 1515 Appendix B: 'Tentative recommended practice to avoid fatigue cracking'. In this the permitted number of cycles is quoted as

$$N = \left\{ \frac{1\,160(3\,000 - T)}{4f_r - 14\,500} \right\}^2 \tag{3.1}$$

where T is the temperature in °C, and f_r is the fatigue stress range in lbf/in^2. The value of f_r is based on the nominal hoop stress. Stress concentrations due to penetrations, etc. are allowed for in the code, but a corrosion allowance is deducted from the actual vessel thickness. For this particular vessel, the internal diameter is 72in (1.83 m) and the net thickness 2·75in (69·9 mm), so $f_r = 13·614P_r$, where P_r is the cyclic pressure range. (The use of stress range rather than absolute values implies, for example, that a loading of 1 400–2 000 lbf/in^2 causes the same damage as one of 400–1 000 lbf/in^2.) For $T = 50°C$ and the fluctuating loading, equation (3.1) gives $N = 35\,430$. As this was greater than 21 900, the manufacturer had concluded, in accordance with Appendix B, that the fatigue strength of the vessel was satisfactory.

Where more than one loading is involved, Appendix B requires that a Miner's rule summation (Section 2.4) in the form

$$\sum \frac{n_i}{N_i} \leqslant 1.0 \tag{3.2}$$

be satisfied, where n_i is the expected number of cycles for the ith load level, and N_i the permitted number of cycles for the ith load level calculated using equation (3.1). For the fluctuating loading, $n_i/N_i = 0.423$, but for the full pressure cycles it is 1.42, showing immediately that the vessel did not meet the requirements of Appendix B. With the expected loading, the vessel would have had a safe life of only 9.6 years, which was not acceptable. However, keeping the fluctuating pressure range (600 lbf/in^2, 41.4 bar) the same, but reducing the maximum working pressure, provided a potentially acceptable solution. Calculations using equation (3.1) and expression (3.2) showed that the desired 15 year design life could be achieved by reducing the maximum working pressure to 1 499 lbf/in^2 (103 bar).

At the time the assessment was carried out, BS 5500 (which will eventually supersede BS 1515) had just been published, and its Appendix C, 'Recommended practice for assessment of vessels subject to fatigue', contains a simplified procedure which is deliberately intended to be very conservative. Calculations to Appendix C indicated that the maximum working pressure, keeping the same fluctuating pressure range as before, should be 75.7 bar. Detail differences in the general requirements of the two standards mean that this is not directly comparable with the higher figure obtained from BS 1515 to which the vessel was designed. Nevertheless, it was decided that the maximum working pressure should be limited to 76 bar, and also that records be kept so that the fatigue life could be reassessed in the light of actual rather than predicted usage. The reduced maximum working pressure was not a serious operating constraint, and the records kept showed that actual usage was less severe than predicted, so no fatigue problems are anticipated in the life of the vessel.

Both BS 1515 and BS 5500 are undoubtedly conservative in their fatigue requirements, and the vessel would almost certainly have withstood the original specified loading. Both standards allow the fatigue clauses to be disregarded provided that evidence of adequate fatigue life, such as more detailed calculations or practical experience of usage, is available. Details of the methods to be used are not specified in the standards and are left to agreement between the manufacturer and the purchaser. Such escape clauses are common in standards, but are primarily intended to cover cases where the necessary evidence is readily available, which was not the case for this particular vessel.

3.5 SOME LEGAL ASPECTS

When a fatigue failure occurs, anyone connected with the design, manufacture, supply, servicing, repair or operation of the product might find themselves in the position of being held legally responsible for personal injury and material damage caused by the failure.

In general, the law on 'product liability' is complex, varies widely from country to country and is undergoing reform in a number of countries.[43] Reference 44 gives a useful outline of the situation, and standard texts on British[45] and American[46] practice are available. Unfortunately, no precise, universally accepted definitions exist of the various legal terms involved, such as 'strict liability'. For criminal liability to be established, i.e. for a person to be held to have committed a criminal offence for which he may be punished, it is usually necessary to prove that he was negligent in some way in his association with the product conceived. For civil liability to be established, i.e. for a manufacturer to have to pay compensation to a person who is injured or suffers loss, it is sometimes necessary to prove negligence. Under 'strict liability' it is merely necessary to show that the product was 'defective', but some defences are possible, whereas under 'absolute' or 'no-fault' liability there is no defence. There is nothing new in the concept of strict liability, but it is being extended to more situations, especially in personal injury cases. It is therefore of increasing concern to those concerned with all types of products. It has been claimed[47] that legislation is tending towards the situation where those concerned with a product are guilty until proved innocent, a reversal of the normal legal principle, and there is growing concern for their position.[48] It is perhaps unfortunate that current legal trends are tending to inhibit innovation.

The legal meaning of 'defective' is obviously of crucial importance, but an unambiguous definition has not emerged. One proposed definition[44] reads:

> A product should be regarded as defective if, at the time when it is put into circulation by whoever is responsible for it as its producer, it does not comply with the standard of reasonable safety that a person is entitled

to expect of it; the standard of safety should be determined objectively having regard to all the circumstances in which the product has been put into circulation, including, in particular, any instructions or warnings that accompany the product when it is put into circulation, and the use or uses to which it would be reasonable for the product to be put in these circumstances.

Clearly this, and similar definitions, include catastrophic fatigue failures which occur during normal use, and is obviously acceptable from the consumer viewpoint. However, from the viewpoint of those responsible for a product it is unsatisfactory in that it could be very difficult indeed to decide in advance whether a particular product might be regarded as defective. Misuse of a product which is not defective in the legal sense can also lead to legal liability. For example, the operator of a fairground ride who deliberately exceeded the manufacturer's maximum recommended speed had to pay compensation when this resulted in a catastrophic fatigue failure.[49]

In the special case of fatigue, it can be helpful to differentiate between design defects (e.g. the calculation described in the previous section) and manufacturing defects (e.g. faulty welding) where the designer's intentions have not been followed during manufacture. Once a fatigue failure has occurred it is always possible to point to some action which would have prevented the failure. However, it is difficult to reconcile the essentially random nature of fatigue (Chapter 5) with any definition of 'defective' which requires precise prediction of future events. Perhaps the only useful general advice that can be given is to be aware of fatigue as a potential problem, and to be clear in one's own mind on the approach (Section 3.3) that is being used to reduce the possibility of fatigue failure to an acceptable level.

Standards were originally written to provide common ground among experts, and it was tacitly understood that expert knowledge was required for their interpretation and use. Increasingly, standards are regarded as quasi-legal documents which are likely to be cited during legal proceedings including product liability litigation, and whose interpretation should therefore not be open to doubt. In response to this trend, the British Standards Institution has issued much stricter guidelines[50] for the writing of standards.

REFERENCES

1 NORWEGIAN PUBLIC REPORT: 'The *Alexander L. Kielland* accident' (in Norwegian with extended summary in English), 1981, Oslo, Bergen and Tronsø, Universitetsforlaget
2 A. COTTRILL: *Offshore Eng.*, 1980, May, 9–10

3 ANON: *Offshore Eng.* 1981, April, 15–16

4 N. E. FROST: *J. Soc. Environ. Eng.*, 1975, **14-2**, (65), 21–24, 27–28

5 R. HALMSHAW: *Met. Eng. Q.*, 1970, **10**, (4), 12–18

6 R. R. WHYTE: 'Engineering progress through trouble', 1975, London, Institution of Mechanical Engineers

7 T. R. GURNEY: 'Fatigue of welded structures', 2nd ed., 1979, London, Cambridge University Press

8 G. G. CHELL: (ed.): 'Developments in fracture mechanics—I', 1979, London, Applied Science Publishers Ltd

9 A. G. PUGSLEY: 'The safety of structures', 1966, London, Edward Arnold

10 O. E. LISSNER: *J. S. Afr. Inst. Min. Metall.*, 1967, **67**, (6), 273–316

11 J. E. FIELD and D. SCOTT: *Proc. Inst. Mech. Eng.*, 1969–70, **184**, (3B), 119–131

12 Proc. Int. Conf. on Fatigue of Metals, 1956, London and New York; 1958, London, Institution of Mechanical Engineers

13 G. M. BOYD: 'Brittle fracture in steel structures', 1970, London, Butterworth

14 R. H. PARSONS: 'History of the Institution of Mechanical Engineers 1847–1947', 1947, London, Institution of Mechanical Engineers.

15 BRITISH STANDARDS INSTITUTION: 'Fusion welding of steel castings. Part I: Production, rectification and repair', BS 4570: Part I: 1970

16 D. BIRCHON: 'Non-destructive testing', Engineering Design Guides 09, 1975, London, Oxford University Press

17 V. F. BIGNELL, G. PETERS and C. PYM: 'Catastrophic failures', 1977, Milton Keynes, Open University Press

18 W. G. HEATH: *Aeronaut. J.*, 1980, **84**, (831), 81–92

19 L. E. STEELE and K. E. STAHLKOPF (eds.): 'Assuring structural integrity of steel reactor pressure vessels', 1980, London, Applied Science Publishers Ltd

20 P. E. JORDAN: *Metallurgia*, 1980, **47**, (7), 364–369

21 H. O. FUCHS: *J. Mech. Des.*, 1980, **102**, (1), 1

22 BRITISH STANDARDS INSTITUTION: 'Steel, concrete and composite bridges. Part 10: Code of practice for fatigue', BS 5400: Part 10: 1980

23 R. A. SMITH (ed.): 'Fracture mechanics: current status, future prospects', 1979, Oxford, Pergamon Press

24 H. O. FUCHS and R. I. STEPHENS: 'Metal fatigue in engineering', 1980, New York, Wiley–Interscience

25 W. SCHÜTZ: *Eng. Fract. Mech.*, 1979, **11**, (2), 405–421

26 H. D. GERLACH: *Int. J. Pressure Vessels Piping*, 1980, **8**, (4), 283–302

27 J. KALLABY and J. B. PRICE: *J. Pet. Technol.*, 1978, **30**, (3), 357–366

28   K. J. MARSH: *J. Soc. Environ. Eng.*, 1974, **13-4**, (63), 15–16, 21–22

29   K. J. MARSH: *Int. J. Fatigue*, 1979, **1**, (1), 3–6

30   W. T. KIRKBY, P. J. E. FORSYTH and R. D. J. MAXWELL: *Aeronaut. J.*, 1980, **84**, (829), 1–12

31   B. J. DABELL and P. WATSON: Int. Conf. on Fatigue Testing and Design, 1976, City University, London; Vol. I, 2.1–2.50, 1976, Buntingford, Herts., Society of Environmental Engineers Fatigue Group

32   D. K. EWING: *Int. J. Fatigue*, 1979, **1**, (2), 89–92

33   L. P. POOK: *J. Soc. Environ. Eng.*, 1978, **17-1**, (76), 22–23, 25–28, 31–35

34   'FALSTAFF (Description of a Fighter Aircraft Loading STAndard For Fatigue evaluation)', 1976, Darmstadt, Laboratorium für Betriebsfestigkeit

35   DEPARTMENT OF ENERGY: 'Descriptions of UK Offshore Steel Research Project. Objective and current progress', Reference MAP 015G(1)/20-2, 1974, London, Department of Energy

36   J. G. WYLDE and A. McDONALD: *Int. J. Fatigue*, 1980, **2**, (1), 31–36

37   Proc. European Offshore Steels Seminar, 1978; 1980, Cambridge, The Welding Institute

38   A. J. TROUGHTON: *J. R. Aeronaut. Soc.*, 1960, **69**, (599), 653–667

39   'Airplane damage tolerance requirements', Military specification MIL-A-8344, 1974, United States Air Force

40   ANON: NEL Report No. 665, 1979, East Kilbride, Glasgow, National Engineering Laboratory

41   BRITISH STANDARDS INSTITUTION: 'Welded pressure vessels (advanced design and construction) for use in the chemical, petroleum and allied industries. Part I: Carbon and ferritic alloy steels', BS 1515: Part I: 1965

42   BRITISH STANDARDS INSTITUTION: 'Specification for unfired fusion welded pressure vessels', BS 5500: 1976

43   Proc. BSSM/I. Prod. E. 1980 Int. Conf. on Product Liability and Reliability, 1980, Newcastle upon Tyne, British Society for Strain Measurement

44   M. HEALY: *ibid.*

45   C. J. MILLER and P. A. LOVELL: 'Product liability', 1977, London, Butterworth

46   L. R. FRUMER and M. I. FRIEDMAN: 'Products liability' (4 vols.), 1975 (with supplement to 1979), New York, Matthew Berder

47   J. NOON: *Design*, 1979, (369), 54–56

48   D. H. HOUSEMAN: as Ref. 43

49   C. ALLEN-IVEY: as Ref. 43

50   BRITISH STANDARDS INSTITUTION: 'General principles of standardization', BS 0: Part 1: 1981

CHAPTER 4

Fatigue crack growth under constant amplitude loading

There is a crack in everything God hath made.
R. W. Emerson, *Essays*, First series.

4.1 INTRODUCTION

Because fatigue crack propagation is of major importance from a practical engineering viewpoint (Chapter 3), the remainder of this book concentrates upon this feature rather than fatigue crack initiation. Those applied mechanics concepts needed for its discussion are included in this chapter for completeness. The applied mechanics framework required for the discussion of the behaviour of crack bodies under load has become known as 'fracture mechanics'. As in all branches of applied mechanics, considerable simplification results if linearity is assumed. In particular, the concept of the stress intensity factor is central to the theory of linear elastic fracture mechanics, and provides a reasonably rigorous framework for the discussion of various aspects of fatigue crack growth. This chapter concentrates on the practical application of the stress intensity factor to constant amplitude fatigue crack growth; various limitations and modifications which arise from the actual behaviour of metallic materials are discussed.

Fracture processes may be viewed at a range of scales,[1] as shown schematically in Fig. 4.1. For comparison, fatigue crack growth rates of practical interest range from about 10^{-8} to 10^{-2} mm/cycle.

Present-day fracture mechanics, which forms the basis of mechanical descriptions (Section 2.1), deals largely with macroscopic aspects of crack behaviour at scales of 10^{-1} mm and above. Fracture surfaces are therefore assumed to be smooth, although on a smaller scale they are generally very irregular. Figure 4.2 shows examples[2] of actual fatigue crack profiles. The basic assumptions are the same as in ordinary strength of materials theory: the material is a homogeneous isotropic continuum (i.e. microscopic irregularities in structure are neglected); stress is proportional to strain; strains are

45

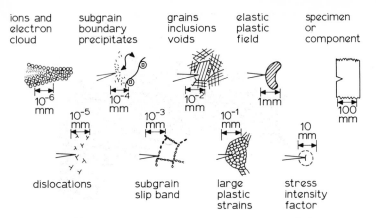

Fig. 4.1   **Schematic illustrations of fracture at different scales under essentially elastic conditions (from McClintock and Irwin[1])**

small; and distortions are neglected. For convenience, the material is also usually assumed to be free from large-scale self-equilibrating internal stress and body forces such as those due to gravity. Metallurgical descriptions (Section 2.1) of the mechanisms[3] are usually at scales of the same order as metallurgical features, i.e. 10^{-4}–10^{-2} mm, and are often referred to as micromechanisms. They are normally qualitative rather than quantitative in nature, but do add to the understanding of relationships between metallurgical features and crack growth rates.

Well established topics are not referenced; for further information see e.g. Refs. 2–9. More recent developments and some specific points are referenced in detail.

4.2   MODES OF CRACK SURFACE DISPLACEMENT

There are three modes of crack surface displacement which can occur when a cracked body is loaded, and in which crack growth can take place. These are (Fig. 4.3):

    I   The opening mode. The crack surfaces move directly apart.

    II   The edge-sliding mode. The crack surfaces move normal to the crack front and remain in the crack plane.

   III   The shear mode. The crack surfaces move parallel to the crack front and remain in the crack plane.

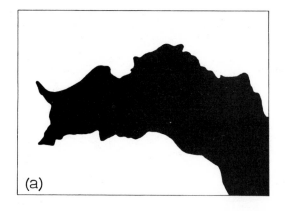

Fig. 4.2 Profile of a crack tip in a loaded mild-steel sheet (from Frost, Marsh and Pook[2]) [×125]

The superimposition of these three modes is sufficient to describe the most general case; it is usual to add the Roman numerals, I, II and III as subscripts to symbols to describe each mode, but other notations are also in current use.

Viewed macroscopically, cracks in isotropic materials under essentially elastic conditions tend to grow in Mode I under both static and fatigue loadings, irrespective of their initial orientation (Section 7.1). Attention in this chapter is therefore largely confined to this mode. At finer scales, all three modes will normally be present. The main exceptions to Mode I growth are when a crack follows a plane of weakness or when a crack grows on planes at 45° through the

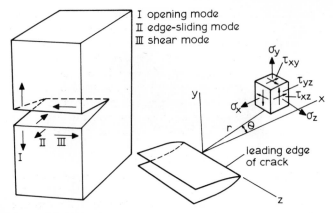

Fig. 4.3   Mode direction notation and crack tip coordinates

thickness. Such slant crack growth usually occurs in thin sheets (Fig. 4.4) and is sometimes called a shear mode, because it is on planes of maximum shear stress in an uncracked sheet. However, this description is erroneous; it is actually a combination of Modes I and III. The direction and plane of growth of a Mode I crack are approximately (exactly for symmetrical configurations) perpendicular to the maximum principal applied stress. A crack which has grown entirely in Mode I is not necessarily straight and crack trajectories are not readily determined (Section 7.5). As a general rule, a crack tends to be attracted by the nearest free surface and may follow a curved path even under initially symmetrical loading conditions. Cracks in structures are frequently found to follow complex paths. Ensuring that a crack maintains its initial plane and direction is an important factor in the design of fracture-mechanics-based test specimens.

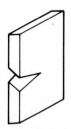

Fig. 4.4   Crack on a 45° plane

The complete solution of a crack growth problem obviously includes the determination of the crack trajectory. However, a crack loaded in Mode I will usually grow in Mode I without change of direction. Under mixed mode loading, the crack direction normally changes abruptly as growth starts (Section 7.3). In basic discussions of the conditions under which a crack will grow, it is usual to consider the two-dimensional case and to assume that the initial crack is straight and that crack growth is an extension of the original crack. This can usually be done without significant loss of generality.

4.3 STRESS INTENSITY FACTORS

Crack surfaces are stress-free boundaries adjacent to the crack tip and therefore dominate the distribution of stresses in that area. Remote boundaries and loading forces affect only the intensity of the stress field at the crack tip. There is a particular type of elastic stress field corresponding to each mode of crack surface displacement; these may be described in terms of the stress intensity factor K, which provides a convenient single parameter description of the crack tip stress field. It has the dimensions (stress) $\times$ (length)$^{1/2}$ and is usually in units of $MN\,m^{-3/2}$. The stress intensity factor actually represents the first term in the series expansion for the elastic stress field at a crack tip, and is, for many purposes, a sufficiently accurate representation of the elastic stress field. Individual stress components (Fig. 4.3) and displacements are proportional to $K/r^{1/2}$, where r is the distance from the crack tip (see Appendix I). Stresses tend to infinity as the crack tip is approached, so K is a mathematical singularity. For sheets of infinite extent, containing a central crack length $2a$, the stress intensity factors are given by

$$K_I = S(\pi a)^{1/2} \qquad K_{II} = \tau(\pi a)^{1/2} \qquad K_{III} = \tau(\pi a)^{1/2} \qquad (4.1)$$

where S is the direct stress and τ is the shear stress on the crack plane. The conventional $\pi^{1/2}$ included in equation (4.1) is sometimes omitted, particularly in early work. For Modes I and II the sheet can be two-dimensional, for Mode III it is assumed to be infinitely thick.

The elastic stress analysis assumes an ideal elastic material, i.e. a perfectly elastic isotropic continuum. For a 'long' crack, the scale of the crack tip region where K represents stresses is 1–10 mm. Any crack tip feature on a smaller scale can be regarded as within the region and does not affect its general character. Practical materials are not homogeneous and yield in the high stress region at a crack tip. Provided that yielding is confined to a small region right at the crack tip, the elastic stresses outside this region are only slightly affected, and the stress intensity factor still provides a reasonable description of the crack tip stress field. Small-scale, non-linear effects due to microstructural irregularities, internal stresses, local irregularities in the crack surface and the actual fracture process can similarly be regarded as within the crack tip stress fields and therefore neglected in a reasonable approximation. It is conventional to treat slant crack growth in thin specimens and the shear lips which often appear on thick specimens (Section 7.2) as Mode I growth when making calculations. Although this is difficult to justify by the 'small-scale' argument, it does not cause difficulties in practice. By a similar argument, stress intensity factors can be used to describe elastic stress fields at sharp notches; indeed, elastic stress fields at the tips of sharp notches are, with due attention to detail, similar to those for cracks (see Appendix I). An apparent objection to the use of the stress intensity factor approach is the violation of the initial assumption of small strain and negligible distortion even at the crack tip. However, if these conditions are only violated locally, the general character of the stress field in the vicinity of the crack tip remains unchanged.

Material properties in the presence of a crack may be measured in terms of the opening mode stress intensity factor K_I, in the same way as the tensile or fatigue properties of a plain specimen are measured in terms of stress. For example, the higher the value of K_I, the more severe the crack, and when a critical value K_c is reached, a crack will extend under a static load; K_c is therefore a measure of a material's fracture toughness, or resistance to brittle fracture. Fatigue crack growth data may conveniently be correlated in terms of ΔK_I, the range of K_I during a fatigue cycle, but neglecting any compressive loads (Section 4.5.1). The use of stress intensity factors to analyse crack growth data has the advantage that behaviour can be predicted for any cracked body configuration for which a stress intensity factor expression is available; a designer is not limited to situations similar to those used to generate the original data.

Values of K_I for some simple finite geometries are given in Fig. 4.5. Many solutions for various configurations have been published; Ref. 10 is a convenient compilation. Stress intensity factors for different loadings on the same body may be combined by algebraic addition, with the overriding proviso that K_I cannot be negative (but see Section 4.5.2). Stress intensity factors can be obtained analytically for only a limited number of cases; more generally, numerical solutions must be obtained using a computer. Published expressions

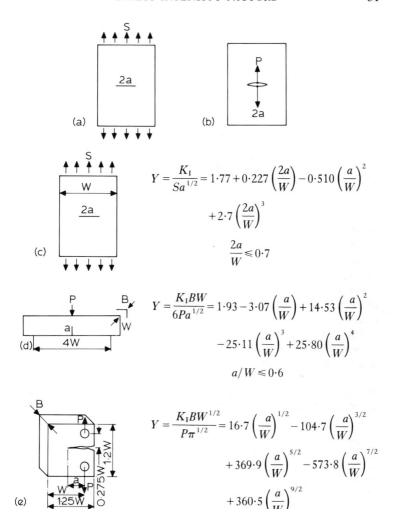

Fig. 4.5 Stress intensity factors for various geometries (from Smith[6]): (*a*) centre crack in infinite plate, remote load, $K_I = S(\pi a)^{1/2}$; (*b*) centre crack in infinite plate, point loads on crack surface, $K_I = P/[B(\pi a)^{1/2}]$; (*c*) centre crack in plate of finite width, remote load; (*d*) edge crack, plate loaded in three-point bend; (*e*) ASTM compact tensile specimen (CTS)

are often for curves fitted to series of discrete results, and care must be exercised not to use the expression outside the ranges specified. For some purposes, expressions can conveniently be written in the general form

$$K_I = S(\pi a)^{1/2} \alpha \qquad (4.2)$$

where α is a geometry correction factor of the order of 1.

Unfortunately, only a limited number of solutions are available for geometries of practical interest such as cracks emanating from notches or fillets, and useful generalizations are not easy. However, empirical fits of engineering accuracy using a gross stress basis have been made[6] for the case of a crack of length l emanating from the root of a semi-elliptical notch, depth d, root radius ρ and $d > \rho$. For $l > 0.13(d\rho)^{1/2}$, the effective crack length (a in equations) is $l + d$, and for $l < 0.13(d\rho)^{1/2}$ it is $l + 7.69l(d/\rho)^{1/2}$. Hydrostatic pressure has no effect on K_I. However, pressure within a crack, not balanced by external pressure, is equivalent to a tensile stress of the same magnitude across the crack.

The effect of introducing a crack into a body containing large-scale residual stresses is to relieve the residual stresses on the crack plane. Provided the crack is not too large, the corresponding stress intensity factors are the same as if stresses, equal in magnitude to the original residual stress but opposite in sign, were applied to the crack surfaces. These must be added algebraically to the stress intensity factors due to external loads. The presence of residual stresses of unknown magnitude can be a serious obstacle to the application of fracture mechanics. Even when residual stresses are known, the calculation of the corresponding stress intensity factors can present difficulties.

4.4 EFFECT OF YIELDING

An uncracked plate loaded in uniaxial tension is in a state of plane stress, i.e. there is no stress in the thickness direction; this is still so in the bulk of the plate when a crack is introduced, but highly stressed material adjacent to the crack tip is constrained by the less highly stressed surrounding material, and stresses are induced in the thickness direction in the interior of the plate at the crack tip.

This situation is referred to as plane strain in fracture mechanics.[2] Limited plastic flow due to yielding of the material adjacent to the crack tip does not affect the situation in a thick plate of a ductile metal (Fig. 4.6): a plate is said to be thick if the thickness is at least $2.5(K_I/S_Y)^2$ (where S_Y = yield stress, usually taken as the 0.2% proof stress). When the thickness is very much less than $2.5(K_I/S_Y)^2$, the

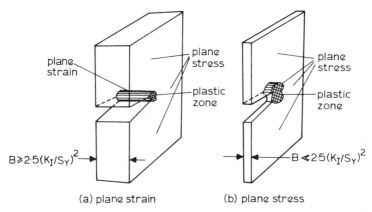

(a) plane strain   (b) plane stress

Fig. 4.6   Stress state, thickness and plastic zone sizes (from Pook[8])

crack tip plastic zone size becomes comparable with the thickness and yielding can take place on 45° planes, relaxing the stresses through the thickness, so that the whole plate is in a state of plane stress. In fatigue, the stress state is conventionally described by the situation at the maximum load in the fatigue cycle.

The approximate radius r_p of the plastic zone at a crack tip in a ductile metal, shown schematically[11] in Fig. 4.7, is given by

$$r_p = \frac{1}{2\pi}(K_I/S_Y)^2 \tag{4.3}$$

for plane stress and one-third of this amount for plane strain. Provided again that r_p is small compared with the crack length and that the maximum net section stress does not exceed $0.8S_Y$, the plastic zone has little effect on the value of K_I; if required, a correction can be made by using (r_p+a) instead of a, in an iterative manner, in calculations. Failure to check whether large-scale yielding might be occurring is the commonest error in the application of linear elastic fracture mechanics, and quoted data for which no check appears to have been made should be regarded as suspect. (Failure to check for yielding can also cause problems in the testing of plain specimens in bending[12] and other situations where the stress field is non-uniform.)

In fatigue, the reversed yielding which takes place on unloading is only of limited extent (Fig. 4.7), so a similar correction to ΔK_I is

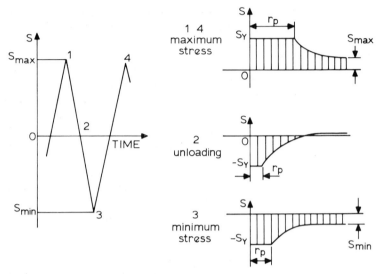

Fig. 4.7   Elastic–plastic stresses at a crack tip under tension–compression loading (from Usami and Shida[11])

not usually necessary. If required, it can be made using equation (4.3) but with K_I replaced by ΔK_I and S_Y by $2S_Y$. The limited extent of the reversed plastic zone means that a plate that appears to be in a state of plane stress may not be so after unloading, a question which does not appear to have been discussed in the literature. Occasionally, the terms plane strain and plane stress are associated with square and slant fracture respectively. This practice has no scientific basis (Section 7.2) and is to be deprecated. The second term of the series expansion for the crack tip stress field (previous section) represents a stress parallel to the crack and hence is strongly influenced by biaxiality. It can have a significant influence on plastic zone size.[5]

4.5   ANALYSIS OF EXPERIMENTAL DATA

The term fatigue crack growth, used without qualification, generally implies macrocrack growth (Section 2.3), i.e. Stage II in Forsyth's notation.[13] Very large numbers of Stage II fatigue crack growth tests have been carried out under constant amplitude loading, i.e. a loading sequence in which all stress cycles are identical. Normally, loads are applied and removed smoothly at moderate frequency, typically around 30 Hz, so that impact conditions and strain rate effects are absent. Most tests are carried out at room temperature in air, and time-dependent effects, such as corrosion, often make little contribution to the growth process. Stage II fatigue crack growth is sometimes taken as synonymous with Mode I fatigue crack growth. This is an oversimplification[2] which neglects the transition to slant crack growth (Fig. 4.4) that sometimes takes place in thin sheets (Section 4.2) after an initial period of Mode I growth. The transition usually occurs when a critical value of ΔK_I for a given material and thickness is exceeded (Section 7.2). Growth rates and crack openings are then relatively large, and the usual striations on the fracture face tend to be replaced by ductile dimples indicating that considerable ductile strain is occurring.[5] In this section, the transition is ignored; it can, however, lead to a thickness dependence in fatigue crack growth behaviour (Section 7.2).

4.5.1 Acquisition of data

Several different ways of presenting crack growth data and 'laws' of fatigue crack growth have been suggested; some are empirical, others have been derived theoretically by making various plausible assumptions. All these laws, some of which have been discussed elsewhere,[2,14] can be regarded as valid in the sense that they describe a particular set of data, and can be used to predict crack growth rates in situations similar to those used to collect the data. It is sometimes possible to fit the same set of data to apparently contradictory laws, and hence no decision on which law is the more correct is possible.[2] In early papers, such as Ref. 15, data were usually presented directly by plotting crack length a against the number of cycles N. Later, various indirect procedures involving plots of da/dN against some function of the applied stress cycle or a, or both, were used. No method was noticeably a better representation of the data than any other, and hence the choice of a particular method of analysis became a matter of the convenience with which the data could be subsequently applied to practical problems. On these grounds (Section 4.3), the use of the stress intensity factor has no serious rivals for the analysis of fatigue crack growth data, and its use, except for certain special situations, is now universal.

The first accurate fatigue crack growth data[2] were obtained from wide, long, thin sheets containing a small transverse central slit to initiate cracks (Fig. 4.5c); the sheet was usually subjected to a wholly tensile loading cycle applied normal to the crack, namely $S_m \pm S_a$

where $S_m > S_a$. This centre-cracked tension specimen is still widely used. The compact tension specimen[16] (Fig. 4.5e) is particularly convenient as it is economical in both material and machine load requirements. When a centre crack or compact tension specimen is used, both K_{Imax} and ΔK_I increase with crack length, but the stress ratio $R = S_{min}/S_{max}$ remains constant ('max' and 'min' refer to the maximum and minimum in a fatigue cycle). For some purposes, a 'constant K' specimen in which the stress intensity factor is independent of crack length is convenient. Various constant K specimens are described in Ref. 17.

Rates of growth are obtained by measuring the crack length at convenient intervals, either manually or by automatic methods,[18] fitting a curve of crack length against number of cycles to the experimental points (Fig. 4.8 shows typical data), and taking the slope of this curve at stipulated crack lengths or at stipulated numbers of cycles. Full accounts of experimental techniques and data reduction methods may be found in a standard test method,[16] and a draft standard which has been circulated for comment.[19] Fatigue crack growth data can also be obtained by measurement of striation spacing on the fracture surface[2,20] or by consideration of the total lives of cracked specimens after testing.[21] Data are usually presented as plots of da/dN against ΔK_I for constant R. If K_{Imax} is well below K_c, ΔK_I plays the major role in determining crack growth behaviour; if

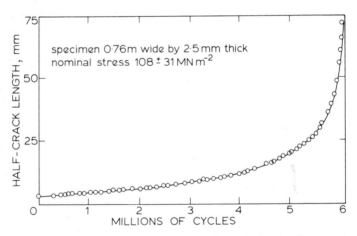

Fig. 4.8   Growth curve for a central crack in a mild steel specimen (from Chell[5])

ΔK_{I} is maintained constant, the growth rate will be constant. Data obtained from specimens containing large-scale residual stresses are identical with those obtained from stress-free specimens provided that the residual stresses are taken into account in the calculation of stress intensity factors.[22] Unsuspected residual stresses, or those of unknown magnitude, can be a serious obstacle to the acquisition of satisfactory data.[22,23]

For many metals, data can be represented by the Paris equation[24]

$$\mathrm{d}a/\mathrm{d}N = D(\Delta K_{\mathrm{I}})^{m} \tag{4.4}$$

where D and m are material constants; m is usually around 3. More elaborate expressions may be used when a better fit to data is required.[14] Results for some materials, such as mild steel, are largely independent of the value of R. Others, such as high strength aluminium alloys, show considerable sensitivity to its value; these are usually referred to as mean stress sensitive materials. In some materials and circumstances, fatigue crack growth rates are thickness dependent.[2] Some typical data[5] are shown in Fig. 4.9. They were obtained on 0.76 m wide $\times 2.5$ mm thick mild steel sheet specimens having central cracks of overall lengths up to 100 mm, and include the data shown in Fig. 4.8. The results are plotted as a scatter band of $\mathrm{d}a/\mathrm{d}N$ against ΔK_{I}. It is seen in general that the data are indeed well represented by equation (4.4), where m is around 3.

At the upper end of a growth rate curve, the growth rate tends to increase beyond that given by the Paris equation as either K_{Imax} approaches K_{c} or $(S_{\mathrm{m}} + S_{\mathrm{a}})$ approaches the yield stress of the material. At the lower end, there is a cut-off at a growth rate of about one lattice spacing per cycle. This is the threshold for fatigue crack growth (Section 2.3), denoted by ΔK_{th}; it is not necessarily sharply defined. Lower average rates are sometimes observed, but crack growth cannot then be taking place along the whole of the crack front on each cycle; there is always an inflection in the $\mathrm{d}a/\mathrm{d}N$ against ΔK_{I} curve at about one lattice spacing per cycle, reflecting the change from continuum to quantum mechanics. Figure 4.10 shows some typical data[25] for the threshold region.

Fatigue crack growth data for various materials,[6] including the mild steel data shown in Fig. 4.9, are shown in Table 4.1 in terms of the exponent m in the Paris equation and the value of ΔK_{I} for a crack growth rate of 10^{-6} mm/cycle. Note that the fatigue crack growth behaviour of a steel tends to be independent of its tensile strength and composition. The use of fatigue crack growth rate data for both material selection[26] and design purposes[27] is well established.

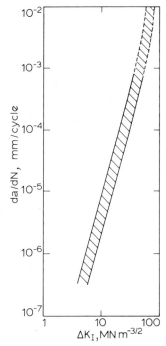

Fig. 4.9   Crack growth rate data for mild steel (from Chell[5])

4.5.2   Plasticity effects

The assumption that a crack closes when the load falls to zero is a convention which is correct only for the purely elastic case. In practice, because of crack tip plasticity effects, a crack may close below zero load; it is propped open by the reversed plastic zone which develops on unloading (Fig. 4.7). On the other hand, because of the wake of plastically deformed material left by a propagating crack, a crack may close at above the minimum load in the cycle even when this is above zero. This phenomenon is referred to as a crack closure (Section 6.1) and reduces the effective values of ΔK_I to below the conventionally calculated values. In both cases, behaviour is usually sufficiently consistent for conventionally calculated values of ΔK_I to form a satisfactory basis for the correlation of data to be used for engineering purposes.

Strictly, ΔK_I should only be used to correlate fatigue crack growth data if the requirements of Section 4.4 are met at the maximum load

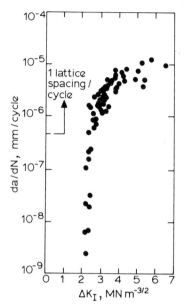

Fig. 4.10   Slow fatigue crack growth data for 7075-T6 aluminium alloy (from Johnson and Paris[25])

in the fatigue cycle. In particular, the maximum net section stress should not exceed 80% of the yield stress, taken as the 0.2% proof stress.[2] (Reference 16 suggests that the net section stress should not exceed the yield stress.) Where the net section stress cannot be defined unambiguously, it may be necessary to adopt an equivalent criterion.[16]

If the maximum cyclic load is such that general yielding occurs, it may still be possible to apply linear fracture mechanics provided that the alternating stress is not too large. In this case, it may be argued that the unloading displacements during the first cycle, and all subsequent cycles, will be essentially elastic so that it is possible to calculate ΔK_I even though K_{Imax} cannot be calculated (see Fig. 4.7). This argument has been found to be justified for some materials where ΔK_I correlates data satisfactorily at stresses beyond general yielding,[2] but such data do not necessarily agree with those obtained under essentially elastic conditions. The broken lines in Fig. 4.9 indicate data of this type. Data should therefore always be checked

Table 4.1 Fatigue crack growth data for various materials

Material	Tensile strength, $MN\,m^{-2}$	0·1 or 0·2% proof stress, $MN\,m^{-2}$	R	m	ΔK_I for $da/dN = 10^{-6}$ mm/cycle, $MN\,m^{-3/2}$
Mild steel	325	230	0·06–0·74	3·3	6·2
Mild steel in brine*	435	—	0·64	3·3	6·2
Cold-rolled mild steel	695	655	0·07–0·43	4·2	7·2
			0·52–0·76	5·5	6·4
			0·75–0·92	6·4	5·2
Low alloy steel*	680		0–0·75	3·3	5·1
Maraging steel*	2 010		0·67	3·0	3·5
18/8 austenitic steel	665	195–255	0·33–0·43	3·1	6·3
Aluminium	125–155	95–125	0·14–0·87	2·9	2·9
5%Mg aluminium alloy	310	180	0·20–0·69	2·7	1·6
HS30W aluminium alloy (1%Mg, 1%Si, 0·7%Mn)	265	180	0·20–0·71	2·6	1·9
HS30WP aluminium alloy (1%Mg, 1%Si, 0·7%Mn)	310	245–280	0·25–0·43	3·9	2·6
			0·50–0·78	4·1	2·15
L71 aluminium alloy (4·5%Cu)	480	415 ⎫	0·14–0·46	3·7	2·4
L73 aluminium alloy (4·5%Cu)	435	370 ⎭	0·50–0·88	4·4	2·1
DTD 687A aluminium alloy (5·5%Zn)	540	495	0·20–0·45	3·7	1·75
			0·50–0·78	4·2	1·8
			0·82–0·94	4·8	1·45
ZW1 magnesium alloy (0·5%Zr)	250	165 ⎫	0	3·35	0·94
AM503 magnesium alloy (1.5%Mn)	200	107 ⎭	0·5	3·35	0·69
			0·67	3·35	0·65
			0·78	3·35	0·57
Copper	215–310	26–215	0·07–0·82	3·9	4·3
Phosphor bronze*	370		0·33–0·74	3·9	4·3
60/40 brass*	325		0–0·33	4·0	6·3
			0·51–0·72	3·9	4·3
Titanium	555	440	0·08–0·94	4·4	3·1
5%Al titanium alloy	835	735	0·17–0·94	3·8	3·4
15%Mo titanium alloy	1 160	995	0·28–0·71	3·5	3·0
			0·81–0·94	4·4	2·75
Nickel*	430		0–0·71	4·0	8·8
Monel*	525		0–0·67	4·0	6·2
Inconel*	650		0–0·71	4·0	8·2

* Data of limited accuracy obtained by an indirect method

to see whether extensive yielding at the maximum load could have affected results.

When the situation cannot be regarded as essentially elastic, elastic plastic fracture mechanics must be used and the situation is much less well defined. The use of ΔJ_I, the range of the Mode I J-integral during the fatigue cycle, has been proposed as a correlating parameter.[5] However, calculation of ΔJ_I is complicated and its use is not widely accepted. The situation is particularly complicated for cracks growing at notches.[28]

4.6 USE OF EXPERIMENTAL DATA

Determination of the fatigue crack growth life of a component subjected to constant amplitude loading, such as the large cast-steel shaft described in Section 3.2.1, is straightforward in principle. The problem is to determine the number of cycles needed for a crack to grow from its initial size a_0 to the final size a_f, at which either the conditions for a brittle fracture are satisfied or the section is so reduced that the working stress exceeds the material's tensile strength. For instance, substituting equation (4.1) in equation (4.4) and integrating gives

$$N = \frac{1/(a_0^{(\frac{1}{2}m-1)}) - 1/(a_f^{(\frac{1}{2}m-1)})}{D \, \Delta S^m \pi^{\frac{1}{2}m}(\frac{1}{2}m - 1)\alpha^m} \tag{4.5}$$

where ΔS is the range of applied stress, but neglecting any compressive stresses. Other crack growth equations and expressions for ΔK_I may be similarly combined, but numerical integration is usually necessary. Care must be taken to ensure that the integration does not extend beyond the ranges of validity of the equations concerned. Provided that this is done, and the crack growth equation used is a reasonable fit to the data, the precise form of the crack growth equation has relatively little influence on the final result.[14] Further expressions which clarify the fatigue behaviour of specimens and structures are easily developed.[29,30]

In the case of a failure analysis, the initial crack size and shape and subsequent pattern of crack growth will be known. However, it may not be possible to calculate accurate stress intensity factors, either because the loading is not known precisely or because appropriate expressions for stress intensity factors are not readily available. Further, appropriate fatigue crack growth rate data may not be available.

For a flaw which has been detected by non-destructive inspection, and whose significance is being assessed, the initial crack size and shape will only be known approximately and the subsequent pattern of crack growth must be estimated. Further uncertainty is introduced where calculations are based on the maximum size of flaw likely to be missed by non-destructive inspection. In addition, allowance must be made for scatter in fatigue crack growth rate data (Section 5.3), by contrast with failure analyses where the use of mean values is appropriate.

The form of equation (4.5) and the shape of crack growth curves obtained in tests (Fig. 4.8) demonstrate that for most of the fatigue crack growth life the crack is relatively short, which makes it difficult to detect fatigue cracks until the growth life is nearly over. This is why fatigue seems to the layman to be an unpredictable phenomenon.

4.7   ANALYSIS AND USE OF THRESHOLD DATA

Values of the fatigue crack growth threshold ΔK_{th} (Sections 2.3 and 4.5.1) may be obtained by a variety of techniques, which all usually give essentially the same result.[31] The most obvious is simply to follow the da/dN against ΔK_I curve downwards. However, unless this is done very carefully (Section 6.4), the threshold can be seriously overestimated.[32] Where the threshold is not sharp, it may be necessary to define it as the value of ΔK_I corresponding to an arbitrary low value of da/dN. A straightforward, if somewhat tedious, technique is to determine an S/N curve for pre-cracked, stress-relieved specimens, with endurance plotted against the initial value of ΔK_I, ΔK_0, rather than stress; the fatigue limit then gives the threshold. Figure 4.11 shows some results[5] for Ni–Cr–Mo–V steel tested at 300°C. Any residual stresses must be allowed for in the measurement of ΔK_{th}, otherwise misleading results may be obtained.[33] For many materials, experimental values of ΔK_{th} correspond reasonably well with values of ΔK_I calculated from the Paris law (equation 4.4) for a value of da/dN equal to one lattice spacing (say, 3×10^{-7} mm).

Values of ΔK_{th} for various materials[34] are shown in Table 4.2. In all cases, data are valid in that the net section stress did not exceed 80% of the yield stress (Section 4.4). Unlike the rate of crack growth, which is largely independent of R, ΔK_{th} normally decreases as R increases. Because of this, a residual stress pattern for which the mean value of K_I falls with crack growth can cause a propagating fatigue crack to slow down and become dormant.[35] Like crack growth rates, thresholds for steels are largely independent of tensile strength or composition, but they can be sensitive to metallurgical factors, especially grain size;[36] higher values are obtained at larger grain sizes. In tests on sharply notched specimens, fatigue cracks formed at the

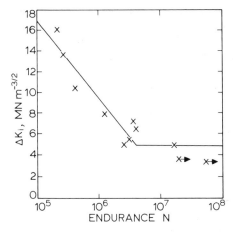

Fig. 4.11   Tests on cracked Ni–Cr–Mo–V steel plates at 300°C (from Chell[5])

notch root sometimes arrest after a short distance. The existence of the threshold explains[2] the occurrence of such 'non-propagating' fatigue cracks (Section 6.4).

For cracks less than about 1 mm in length, it has been found[2,37] that ΔK_{th} is no longer geometry independent, but decreases as crack length decreases. The effect is particularly marked at zero mean stress ($R = -1$). An empirical fit to short-crack threshold data can be obtained by increasing a by an amount l_0 when calculating ΔK_{th}. The value of l_0 depends on the material.[38] For cracks between $2 \cdot 5 \times 10^{-2}$ and 5 mm long, the value of $S_a^3 a$ necessary for crack growth (C) provides a good empirical fit[2] to threshold data obtained at zero mean stress using the method shown in Fig. 4.11. Values of C are included in Table 4.2. The use of C is particularly convenient for calculations, but it can only be used to characterize the threshold where the crack is small compared with other dimensions, and can therefore be regarded as being in a uniform stress field. For example, consider the large cast-steel shaft described in Section 3.2.1. The stress at the failure site was 3 ± 108 MN m^{-2}. As a crack would be closed before the minimum stress is reached,[2] this is equivalent to 0 ± 111 MN m^{-2}. Taking a typical C value for steel of 510 MN m units (Table 4.2) gives the maximum permissible flaw size as 0.37 mm. As this is small compared with the radius (15 mm) at which failure took place, the use of C is satisfactory.

Table 4.2 Values of ΔK_{th} for various materials

Material (ferrous)	Tensile strength, MN m^{-2}	R	ΔK_{th}, MN m$^{-3/2}$	C, MN m units
Mild steel	430	$-1\cdot00$	$6\cdot4$	510
		$0\cdot13$	$6\cdot6$	
		$0\cdot35$	$5\cdot2$	
		$0\cdot49$	$4\cdot3$	
		$0\cdot64$	$3\cdot2$	
		$0\cdot75$	$3\cdot8$	
Mild steel at 300°C	480	$-1\cdot00$	$7\cdot1$	
		$0\cdot23$	$6\cdot0$	
		$0\cdot33$	$5\cdot8$	
Mild steel in brine	430	$-1\cdot00$	$\sim2\cdot0$	
		$0\cdot64$	$1\cdot15$	
Mild steel in brine with cathodic protection	430	$0\cdot64$	$3\cdot9$	
Mild steel in tap water or SAE30 oil	430	$-1\cdot00$	$7\cdot3$	
Low alloy steel	835	$-1\cdot00$	$6\cdot3$	510
	680	$0\cdot00$	$6\cdot6$	
		$0\cdot33$	$5\cdot1$	
		$0\cdot50$	$4\cdot4$	
		$0\cdot64$	$3\cdot3$	
		$0\cdot75$	$2\cdot5$	
Ni–Cr–Mo–V steel at 300°C	560	$-1\cdot00$	$7\cdot1$	
		$0\cdot23$	$5\cdot0$	
		$0\cdot33$	$5\cdot4$	
		$0\cdot64$	$4\cdot9$	
Maraging steel	2 010	$0\cdot67$	$2\cdot7$	
18/8 austenitic steel	685	$-1\cdot00$	$6\cdot0$	540
	665	$0\cdot00$	$6\cdot0$	
		$0\cdot33$	$5\cdot9$	
		$0\cdot62$	$4\cdot6$	
		$0\cdot74$	$4\cdot1$	
Grey cast iron	255	$0\cdot00$	$7\cdot0$	
		$0\cdot50$	$4\cdot5$	
Material (non-ferrous)				
Aluminium	77	$-1\cdot00$	$1\cdot0$	4
		$0\cdot00$	$1\cdot7$	
		$0\cdot33$	$1\cdot4$	
		$0\cdot53$	$1\cdot2$	
L65 aluminium alloy (4·5%Cu)	450	$-1\cdot00$	$2\cdot1$	19
	495	$0\cdot00$	$2\cdot1$	
		$0\cdot33$	$1\cdot7$	
		$0\cdot50$	$1\cdot5$	
		$0\cdot67$	$1\cdot2$	

Table 4.2 (cont.)

Material (non-ferrous)	Tensile strength, $MN\,m^{-2}$	R	ΔK_{th}, $MN\,m^{-3/2}$	C, MN m units
L65 aluminium alloy (4·5%Cu) in brine		0·50	1·15	
ZW1 magnesium alloy (0·6%Zr)	250	0·00	0·83	
		0·67	0·66	
AM503 magnesium alloy (1·6%Mn)	165	0·00	0·99	
		0·67	0·77	
Copper	225	−1·00	2·7	56
	215	0·00	2·5	
		0·33	1·8	
Titanium	540	0·60	2·2	
Nickel	455	−1·00	5·9	700
	430	0·00	7·9	
		0·33	6·5	
		0·57	5·2	
		0·71	3·6	
Monel	525	−1·00	5·6	360
		0·00	7·0	
		0·33	6·5	
	525	0·50	5·2	560
		0·67	3·6	
Inconel	655	−1·00	6·4	750
	650	0·00	7·1	
		0·57	4·7	
		0·71	4·0	

4.8 SHORT-CRACK LIMITATIONS OF STRESS INTENSITY FACTORS

As crack length decreases, the applied mechanics scales at the right-hand side of Fig. 4.1 telescope into the structural feature scales, and it may no longer be possible to regard a structural feature as being contained within the crack tip stress field described by the stress intensity factor (Section 4.3), which is a reasonable approximation at distances not exceeding $a/10$ from the crack tip.[39] Therefore, for a global application of fracture mechanics, the near crack tip field should be at least the size of the largest structural feature of a real material, i.e. the grain size.[6] This size is typically $\sim 10^{-2}$ mm, so a crack would have to be at least 10^{-1} mm long before a stress intensity factor could be used. However, if a crack within a single grain is considered, its total size, $2a$, could not exceed the grain size: the near-tip stress field would extend over a distance of some 5×10^{-4} mm, several orders of magnitude greater than the lattice spacing and comfortably satisfying the continuum requirements. Within a grain, it may no longer be possible to regard the material as isotropic. Fortunately, the concept of the stress intensity factor can be extended to anisotropic materials,[2] so this is not a serious limitation.

Now consider a growing fatigue crack. Rather than using a fracture mechanics equation to calculate a notional plastic zone size, it can be argued that the minimum structural feature in the material associated with the crack tip zone would be a subgrain slip band, $\sim 5 \times 10^{-4}$ mm. It has been suggested[6] that for stress intensity factors to be used the crack size would have to be 50 times larger than such a feature, making the minimum crack size amenable to fracture mechanics treatment using stress intensity factors[39] about $2 \cdot 5 \times 10^{-2}$ mm. Hence, all the values of ΔK_{th} given in Table 4.2 may be regarded as valid.

Now consider the case of a fatigue crack on the point of growing and therefore subjected to the threshold stress intensity factor range ΔK_{th}. As the crack size to which this threshold level is applied becomes smaller, a larger applied stress is required to maintain a constant ΔK_{th}. If, however, this stress exceeds the plain fatigue limit, crack growth is bound to occur. Thus, the threshold for some short cracks is controlled by the plain fatigue limit, causing an apparent decrease in the threshold stress intensity factor. There are two main consequences: firstly, there exists some maximum crack size which does not affect the plain specimen fatigue limit,[2] and secondly, there is some minimum crack size below which stress intensity factors are no longer valid.[40] The changeover from threshold-controlled behaviour to fatigue-limit-controlled behaviour is not sharp, both because of the dependence of ΔK_{th} on crack length (see previous section) for which an R-curve type mechanism has been suggested (Section 6.4), and because of the loss of validity of stress intensity factors. This has been observed in a range of experimental data.[34] Some typical results[41] are shown in Fig. 4.12.

For the practical case of a crack initiated at the root of a notch, the notch-generated plasticity may totally surround the small crack, thus invalidating the stress intensity factor concept.[42] In such cases, the J-integral can be used (Section 4.5), but it is much less convenient.

4.9 MECHANISMS OF FATIGUE CRACK GROWTH
Under fatigue loading a crack can grow even though the value of K_{Imax} corresponding to the range ΔK_I may be much less than K_c.

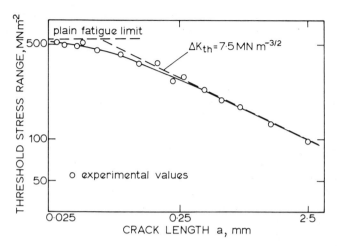

Fig. 4.12 **Effect of crack size on the fatigue crack growth threshold for mild steel, $R = 0$ (from Kitagawa and Takahashi[41])**

In fact, crack growth occurs under fatigue loading not as a consequence of any progressive structural damage, but simply because unloading resharpens the crack tip at each cycle[2] (Section 2.3). The sequence, which is a consequence of crack tip plasticity, is shown schematically in Fig. 2.4. Any satisfactory theory of fatigue crack growth must be based on this type of sequence; in particular, it cannot be based on the accumulation of 'damage' ahead of the crack tip.[2] Fatigue crack growth theories developed so far are based on continuum mechanics and involve deduction of the increment of crack growth in each cycle from an analysis of the changing crack tip geometry during its opening and closing. As a geometry change is involved, any theory of crack growth must use an incremental rather than a deformation theory of plasticity.[43] Any such continuum mechanics theory of crack growth always predicts[2,44] that, for small-scale yielding, the rate of crack growth is proportional to $(\Delta K_{\mathrm{I}}/E)^2$, where E is Young's modulus, unless a characteristic dimension is introduced, or mechanisms fail to remain geometrically similar.[45] It could be argued that such theories are inappropriate because increments of crack growth predicted are on a much finer scale than that at which the material may be regarded as a continuum. However, attempts to develop an atomistic theory of fracture based on quantum mechanics appear to be limited to static loading,[9] and detailed studies of the physics of material behaviour at very fine scales to uncracked bodies.[46]

In the case of, for example, a specimen subjected to a zero to maximum tensile loading cycle, the sequence of crack opening and closing is as shown schematically in Fig. 2.4. Pook and Frost[47] assumed that the part of the crack tip profile calculated from equation (I.1) (see Appendix I) subjected to tensile stresses greater than the yield stress retains its length on unloading, and for ductile metals where E/S_Y is of the order of 10^3, this leads to

$$\frac{da}{dN} = \frac{9 \cdot 3}{\pi} \left(\frac{\Delta K_I}{E} \right)^2 \tag{4.6}$$

It is interesting to note that a limited range of experimental data have been said[48] to conform to

$$\frac{da}{dN} = \frac{8}{\pi} \left(\frac{\Delta K_I}{E} \right)^2 \tag{4.7}$$

Equation (4.6) is for plane stress; for plane strain conditions the equivalent equation, with Poisson's ratio taken as $\frac{1}{3}$, is

$$\frac{da}{dN} = \frac{6 \cdot 1}{\pi} \left(\frac{\Delta K_I}{E} \right)^2 \tag{4.8}$$

which is not greatly different from the plane stress case.* Although the equations were derived for the loading cycle 0 to S, an argument based on crack closure (Section 4.5.2) suggests[47] that they may be applied to other load cycles.

A theory of this type is based on a very much simplified view of the mechanism of fatigue crack growth, which in detail is usually complex.[3] It would be possible to refine the arguments by incorporating crack tip profiles and details of yielding obtained from computer programs, or details such as the yielding on shear bands actually observed at fatigue crack tips.[44] However, if based on idealized material behaviour, such a refined theory would simply alter the numerical factors in equations (4.6) and (4.8). Despite its limitations, it does predict various general features of fatigue crack growth behaviour, especially the dependence on Young's modulus rather

---

* The factor given in Ref. 47 is incorrect.

than strength. Data for materials with a wide range of Young's moduli[49] are compared on a strain basis in Fig. 4.13, where da/dN is plotted against $\Delta K_I/E$. It is seen that the data fall within a band. It follows therefore that the growth rate characteristics of any alloy might be similar to those of the base metal if alloying results in no change in modulus. The data in Table 4.1 show that this is the case for steels, and it is in general true. The theoretical line obtained from equation (4.6) agrees well with experimental rates at around 10^{-5} mm/cycle. However, experimental rates are underestimated at high growth rates and overestimated at low growth rates, both being regions where mechanisms are known to change, with a concomitant change in the numerical factors in equations (4.6) and (4.8). At high rates of crack growth, striations (Fig. 2.2, Section 2.3) tend to give way to ductile dimples or cleavage facets, depending on the ductility

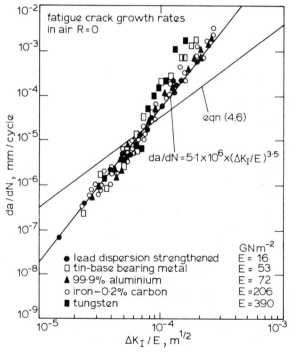

Fig. 4.13 Crack growth rate data for various materials compared on the basis of strain (from Speidel[49])

of the material, and fatigue fracture surfaces are sometimes indistinguishable from those produced by monotonic loading.[5]

At medium rates of crack growth, rates are usually largely independent of a material's tensile strength or composition, but anything that significantly modifies crack tip profiles will affect crack growth rates. For example, in certain titanium alloys, Widmanstätten structures induce crack branching and significantly reduce crack growth rates.[50] As the fatigue crack growth threshold is approached, the behaviour becomes progressively more sensitive to microstructure,[3] and crack growth tends to be confined to crystallographic planes, modifying the crack tip profiles. At high rates of crack growth, ductile dimples, which tend to form ahead of the main crack front, again modify the profiles.

The presence of brittle particles, especially brittle intermetallics, in the microstructure of a material, such as in the high strength aluminium alloys, can result in sudden spurts of crack growth. As the crack tip reaches an intermetallic inclusion and this fractures suddenly, an increment of crack growth whose size depends on the inclusion size is added. Thus, although comparison of striation spacing in different aluminium alloys[51] reveals similar growth rates through the matrix for a given ΔK_I, the overall growth rate is anomalously fast for those alloys containing inclusions which lead to brittle fracture jumps; the effect is illustrated in Fig. 4.14. It also leads to a marked dependence on tensile mean stress, and the presence of such static modes of crack growth can result in thickness dependence in fatigue crack growth behaviour.[2] A less drastic effect is produced by voids,[52] and a somewhat similar effect is sometimes observed in the earlier stages of crack growth when adjacent cracks coalesce.[53]

It can therefore be seen that ΔK_I is a satisfactory parameter to describe crack growth data, but factors such as mean stress, environment, frequency or inhomogeneity of material may affect crack growth rates. However, except in the case of corrosion, such factors do not affect the basic mechanism by which a macrocrack grows, i.e. the continual resharpening of its tip as the load is reduced. It is perhaps convenient to define a neutral environment as one which does not affect the basic mechanism of crack growth, an inert environment as one which permits a degree of rebonding on unloading and therefore retards growth, and a corrosive environment as one which assists crack growth.

For many metals, air can be regarded as a more or less neutral environment, leading to the relatively simple patterns of fatigue crack

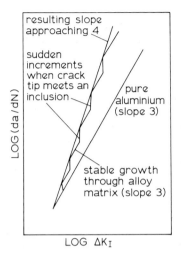

**Fig. 4.14   Non-metallic inclusions giving rise to faster
growth rates (from Smith[6])**

growth behaviour described in this chapter. For both inert environments, such as a vacuum, and corrosive environments, such as
sea-water, behaviour can be much more complicated,[2,5,49] with da/dN
against ΔK_I curves following irregular shapes and a strong frequency
dependence appearing. In corrosive environments, geometry-depen-
dent effects may appear, making it difficult or impossible to apply
results obtained from specimens to practical problems. In particular,
ease of access of the environment to the crack tip must be considered,
and crack closure (Section 4.5.2) may be caused by the deposition
of corrosion products.[5] Compact tension specimens (Fig. 4.5e) are
particularly convenient for corrosion fatigue crack growth studies,
but results must be interpreted with care because, at a given K_I, the
amount of crack opening away from the crack tip is unusually large.[54]

   The numerous factors which affect fatigue crack growth rates mean
that, although ΔK_I is a satisfactory parameter to correlate crack
growth data and provides an applied mechanics framework for the
discussion of the fracture process, there will always be a need to
determine material coefficients experimentally to determine the
effects of mean stress, environment, frequency, inhomogeneity of
material, and other such factors.

REFERENCES

1   F. A. McCLINTOCK and G. R. IRWIN: 'Fracture toughness testing', ASTM STP 381, 84–113, 1967, Philadelphia, Pa., American Society for Testing and Materials

2   N. E. FROST, K. J. MARSH and L. P. POOK: 'Metal fatigue', 1974, Oxford, Clarendon Press

3   J. T. FONG (ed.): 'Fatigue mechanisms', ASTM STP 675, 1979, Philadelphia, Pa., American Society for Testing and Materials

4   D. BROEK: 'Elementary engineering fracture mechanics', 3rd ed., 1982, The Hague, Martinus Nijhoff Publishers

5   G. G. CHELL (ed.): 'Developments in fracture mechanics—I', 1979, London, Applied Science Publishers Ltd

6   R. A. SMITH (ed.): 'Fracture mechanics: current status, future prospects', 1979, Oxford, Pergamon Press

7   T. R. GURNEY: 'Fatigue of welded structures', 2nd ed., 1979, London, Cambridge University Press

8   L. P. POOK: NEL Report No. 465, 1970, East Kilbride, Glasgow, National Engineering Laboratory

9   G. P. CHERAPANOV: 'Mechanics of brittle fracture' (in Russian), 1974, Moscow, Nauka Publishers; English translation with supplementary (1977) material, 1979, New York, McGraw-Hill

10   D. P. ROOKE and D. J. CARTWRIGHT: 'A compendium of stress intensity factors', 1976, London, HMSO

11   S. USAMI and S. SHIDA: *Fatigue Eng. Mater. Struct.*, 1979, **1**, (4), 471–481

12   R. HOLMES and L. P. POOK: NEL Report No. 583, 1974, East Kilbride, Glasgow, National Engineering Laboratory

13   P. J. E. FORSYTH: Proc. of Crack Propagation Symp., 1961; 76–94, 1962, Cranfield, College of Aeronautics

14   D. W. HOEPNER and W. E. KRUPP: *Eng. Fract. Mech.*, 1974, **6**, (1), 47–70

15   A. V. DeFOREST: *J. Appl. Mech.*, 1936, **3**, (1), A23–A25

16   ASTM E647-81: 'Standard test method for constant-load-amplitude fatigue crack-growth rates above 10^{-8} m/cycle', 1981, Philadelphia, Pa., American Society for Testing and Materials

17   S. Ya. YAREMA: *Eng. Fract. Mech.*, 1979, **12**, (3), 365–375

18   C. J. BEEVERS (ed.): 'The measurement of crack length and shape during fracture and fatigue', 1980, Warley, West Midlands, Engineering Materials Advisory Services Ltd

19   ANON: *Sov. Mater. Sci.*, 1979, **15**, (3), 261–275

20   R. C. BATES, W. G. CLARK and D. M. MOON: 'Electron microfractography', ASTM STP 453, 192–214, 1969, Philadelphia, Pa., American Society for Testing and Materials

21   J. M. LOVEGROVE, A. S. SALAH EL DIN and O. K. DAOUD: *Int. J. Fract.*, 1978, **14**, (5), R241–R245

22   J. H. UNDERWOOD, L. P. POOK and J. K. SHARPLES: 'Flaw growth and fracture', ASTM STP 631, 402–415, 1977, Philadelphia, Pa., American Society for Testing and Materials

23   R. J. BUCCI: 'Fracture mechanics' (ed. R. Roberts), ASTM STP 743, 28–47, 1981, Philadelphia, Pa., American Society for Testing and Materials

24   P. C. PARIS and F. ERDOGAN: *Trans. ASME, J. Bas. Eng.*, 1963, **85**, (4), 528–553

25   H. H. JOHNSON and P. C. PARIS: *Eng. Fract. Mech.*, 1968, **1**, (1), 3–45

26   R. J. BUCCI: *ibid.*, 1979, **12**, (3), 407–441

27   L. P. POOK: *Trans. N.E. Coast Inst. Eng. and Shipbuilders*, 1974, **90**, (3), 77–92

28   M. H. EL HADDAD, N. E. DOWLING, T. H. TOPPER and K. W. SMITH: *Int. J. Fract.*, 1980, **16**, (1), 15–30

29   L. P. POOK: *Weld. Res. Int.*, 1974, **4**, (3), 1–24

30   L. P. POOK: NEL Report No. 645, 1977, East Kilbride, Glasgow, National Engineering Laboratory

31   K. JERRAM and E. K. PRIDDLE: *J. Mech. Eng. Sci.*, 1973, **15**, (4), 271–273

32   S. J. HUDAK, A. SAXENA, R. J. BUCCI and R. C. MALCOLM: Report No. AFML-TR-78-40, 1978, Wright–Patterson Air Force Base, Ohio, Air Force Materials Laboratory

33   C. VOSIKOVSKY, L. P. TRUDEAU and J. RIVARD: *Int. J. Fract.*, 1980, **16**, (4), R187–R190

34   L. P. POOK: 'A general introduction to fracture mechanics', 114–135, 1979, London, Mechanical Engineering Publications Ltd

35   P. STANLEY (ed.): 'Fracture mechanics in engineering practice', 1977, London, Applied Science Publishers Ltd

36   C. J. BEEVERS: *Met. Sci.*, 1977, **11**, (8 & 9), 362–367

37   A. ELSENDER, P. HOPKINS and A. D. BATTE: *Met. Technol.*, 1980, **7**, (6), 256–258

38   M. H. EL HADDAD, K. N. SMITH and T. H. TOPPER: *Trans ASME, J. Eng. Mater. Technol.*, 1979, **101**, (1), 42–46

39   J. LANKFORD: *Int. J. Fract.*, 1980, **16**, (1), R7–R9

40   S. J. HUDAK: *Trans ASME, J. Eng. Mater. Technol.*, 1981, **103**, (1), 26–35

41   H. KITAGAWA and S. TAKAHASHI: Proc. of Second Int. Conf. on the Mechanical Behaviour of Materials, 627–631, 1976, Boston

42   R. A. SMITH and K. J. MILLER: *Int. J. Mech. Sci.*, 1978, **20**, (4), 201–206

43   B. BUDIANSKY: *J. Appl. Mech.*, 1959, **26**, (2), 259–264

44   B. TOMKINS: 'Fracture mechanics in design and service', 31–44, 1981, London, Royal Society

45   J. R. RICE: 'Fatigue crack propagation', ASTM STP 415, 247–309, 1967, Philadelphia, Pa., American Society for Testing and Materials

46   J. M. ZIMAN: 'Models of disorder. The theoretical physics of homogeneously disordered systems', 1979, London, Cambridge University Press

47   L. P. POOK and N. E. FROST: *Int. J. Fract.*, 1973, **9**, (1), 53–61

48   P. K. POULOSE, J. E. MORRAL and A. J. McEVILY: 'Prospects of fracture mechanics' (eds. G. C. Sih, H. C. van Elst and D. Broek), 161–177, 1974, Leyden, Noordhoff

49   M. O. SPEIDEL: 'Advances in fracture research, Vol. 6' (ed. D. François), 2685–2704, 1982, Oxford, Pergamon Press

50   G. R. YODER, L. R. COOLEY and T. W. CROOKER: *Eng. Fract. Mech.*, 1979, **11**, (4), 805–816

51   P. J. E. FORSYTH: *Met. Technol.*, 1978, **5**, (10), 351–357

52   R. A. DASQUETA and R. A. QUEENEY: *Int. J. Fatigue*, 1980, **2**, (3), 113–117

53   H. KITAGAWA: 'Fatigue thresholds. Fundamental and engineering applications, Vol. 2' (eds. J. Bäcklund, A. F. Blom and C. J. Beevers), 1051–1068, 1982, Warley, West Midlands, Engineering Materials Advisory Services Ltd

54   B. CROSS, E. ROBERTS and J. E. SRAWLEY: *Int. J. Fract. Mech.*, 1968, **4**, (3), 267–289

CHAPTER 5

Some statistical aspects of fatigue

Of that there is no manner of doubt—
No probable, possible shadow of doubt—
No possible doubt whatever.
W. S. Gilbert, *The Gondoliers*, I.

5.1 INTRODUCTION

In the traditional type of fatigue experiment, some type of loading is applied to a specimen and the number of cycles to failure is recorded. Failure is often taken as complete rupture of the specimen, although other criteria, such as the appearance of a crack of specified size, may be used. Metal fatigue has long been recognized as a random phenomenon, and the consequent scatter in results, even in carefully controlled experiments, complicates both the analysis of experimental data and their subsequent application to engineering problems. The scatter observed in fatigue experiments is greater than is accounted for by inevitable experimental inaccuracies, although these of course contribute to the overall scatter. It is consequently pertinent to examine fatigue from probabilistic viewpoint. From a macroscopic viewpoint it is often convenient to regard a metallic material as a homogeneous continuum, and basing engineering calculations on this assumption does not generally lead to serious error. However, the scatter observed in the fatigue response of a metallic material arises precisely because it is not a homogeneous continuum when considered on a microscopic scale (Section 4.1).

When a S/N curve is to be fitted to individual test points, a great deal of subjective judgment must be exercised. Statistical methods provide a rational approach to this type of problem, but do not of themselves either avoid the need for subjective judgement at some stage, or increase the amount of information present in a given set of data. Statistical methods of analysing scatter in the results of such traditional experiments are well established;[1] standard methods are available.[2,3] Similar methods have been used[4] to derive the design curves given in BS 5400.[5] Statistical theory also provides information on the most efficient use of a limited number of test specimens, and the number of test specimens required to give a specified degree of confidence in the test results.[6,7] The selection of the particular statistical method to be used for a particular application can lead to some subtle problems in scientific method.[8] The approach known as the method of maximum likelihood[9] helps in the selection of methods for the derivation of design curves.

Some of the consequences of the random nature of fatigue are discussed in this chapter. The next section gives mathematical descriptions of some types of fatigue loading, including random loading. Available probability theory,[10,11] which previously has been used largely for random vibration and electronics applications, provides an appropriate mathematical framework for a qualitative discussion of the problem of scatter in fatigue crack growth.[12] Fracture surfaces are generally irregular[13] and can in principle be described using the theory of fractals.[14] Terms such as 'specimen' and 'test' are taken to include both laboratory specimens and the application of service loads to practical structures. Some of the functions involved are, strictly speaking, discontinuous but may be treated as continuous without difficulty.

5.2   MATHEMATICAL DESCRIPTION OF SOME FATIGUE LOADINGS

A proposed definition of metal fatigue, given in Chapter 1, is 'failure of a metal under a repeated or otherwise varying load which never reaches a level sufficient to cause failure in a single application'. This definition is intended to exclude quasi-static tests such as tensile and creep tests, although the boundary between fatigue tests and other tests is not always clear. Mathematically, the loading applied during a fatigue test is an example of a stochastic process[10] and may be defined in terms of a varying force F, which is applied as a function of time $F(T)$ in some time interval. Restricting the varying force to cases where there is at least one maximum and at least one minimum in the intervals considered excludes most quasi-static tests. Where more than one force is applied, they need not be proportional, and only one force need vary and meet the restriction. Fatigue is a strain-controlled process (Sections 2.3 and 4.9), so a fatigue loading can also be defined in terms of a nominal strain ε, chosen to be representative of the strain tensor,[15] at a reference point or region such as the test section of a conventional plain specimen.

In many fatigue applications, behaviour can be considered to be linear elastic[1] so that the stress S at the test section is proportional to F and ε, and a fatigue loading may then be defined in terms of the time function $S(T)$. In a constant amplitude loading, individual cycles can be distinguished and S varies in a regular cyclic manner, for example

$$S(T) = S_m + S_a \sin 2\pi fT \tag{5.1}$$

where S_m is mean stress, S_a alternating stress, f test frequency, and T time. The process is statistically stationary in that the coefficients S_m, S_a and f are unaffected by a shift of the origin along the time axis.

As the process is cyclic, $S(T)$ can be replaced by a continuous function of the number of cycles N, $S(N)$, where $N = T/f$. In fatigue, it is maxima and minima and number of cycles (rather than time) which are usually the main controlling parameters; the shape of the intervening curve between a maximum and minimum is of little importance. It is usual to regard a cycle as the interval between successive minima, to consider only integral numbers of cycles, and to write $S(N)$ as the discontinuous function

$$S(N) = S_m \pm S_a \tag{5.2}$$

which is understood to give successive maxima and minima. Under variable amplitude loading $S(N)$ varies with N. In a programme (or block) loading, $S(N)$ varies in a stepwise fashion, and it may be necessary to take note of part cycles at the changeover between blocks. Figure 5.1 shows a simple two-level block loading.

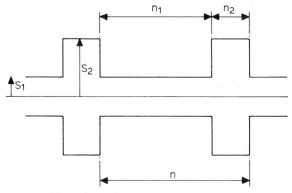

Fig. 5.1   A two-level block loading

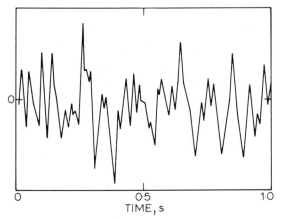

Fig. 5.2   Broad-band random loading (from Chell[16])

5.2.1   Random loadings

A random loading is one in which $S(T)$ is a (usually statistically stationary) random function. Characterization using random process theory is usually straightforward.[11] In the general case of a broad-band random loading[16] (Fig. 5.2), individual cycles cannot be distinguished, and this can make the analysis of test results difficult.[17] Various measures of band width are available. In fatigue, the irregularity factor, which is the ratio of mean level crossings to peaks, is often used;[18] it is less than 1 for broad-band loading. Figure 3.4 is an example of a non-stationary broad-band random loading; characterization of such non-stationary processes, for which statistical parameters vary with time, can be difficult.[11] A narrow-band random process (Fig. 5.3) results when a broad-band random signal is applied to a sharply tuned resonant system.[11,16] Individual sinusoidal cycles

Fig. 5.3   Narrow-band random loading (from Chell[16])

appear, whose frequency corresponds to the resonant frequency of the system; they have a slowly varying random amplitude, and the irregularity factor approaches unity. There is no precise generally agreed definition of what constitutes narrow-band random loading,[18] although it is sometimes taken as 1 where the irregularity factor is at least 0·99. For practical test work, non-stationary processes may be simulated by arranging different levels of stationary processes in blocks. Figure 3.5 shows an example.

In conventional random signal analysis,[11] a first order average is used to give the mean process strength, and a second order average to give the root mean square (rms) value σ of the process. In fatigue, this is usually calculated from the mean value of the process, and is therefore a standard deviation in statistics terminology.[10] Other conventions are sometimes used in the calculation of σ, especially when computer-generated processes are involved,[19] and care is needed when comparing data. The rms gives some indication of the variability of the process and may in addition, in some applications, have a physical significance (e.g. the heating effect of an electric current). For fatigue, the mean may be important, but the rms has no physical significance.

As an alternative to rms, the use of the mth root of the mean mth power, where m is the exponent in equation (4.4), has physical significance in some situations and facilitates data comparison.[20] Appropriate definition of an mth order average of K_I, K_h, leads to $K_h = \Delta K_I$ for constant amplitude fatigue crack growth. This approach has been successfully used for both narrow and broad random loadings.

The form of a random process is usually characterized by the probability distribution of individual values and the power spectral density, which is a measure of the frequency content, or band width, of the process. The irregularity factor can be derived from the power spectral density[18] but, although it is conceptually simpler, much useful information is lost. Probability distributions of random wave forms may be described by the exceedance function $P(S/\sigma)$, which is the probability that the magnitude of S (normalized by dividing by σ) will exceed a certain value. The cumulative probability is $1 - P(S/\sigma)$, and the probability density $p(S/\sigma)$ is the differential of $1 - P(S/\sigma)$. It can therefore be seen that $P(S/\sigma)$ is equal to the area under the curve of $p(S/\sigma)$ versus S/σ between S/σ and ∞.

The two distributions most often encountered in fatigue work[1] are the Gaussian or normal distribution and the Rayleigh distribution. The Gaussian probability density function, shown in Fig. 5.4a,

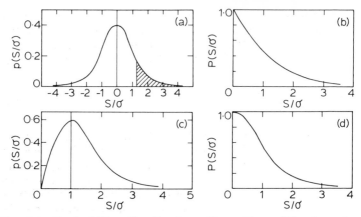

Fig. 5.4   **Probability distributions (from Frost, Marsh and Pook[1]): (a) Gaussian probability density; (b) Gaussian exceedance function; (c) Rayleigh probability density; (d) Rayleigh exceedance function**

is given by the equation

$$p\left(\frac{S}{\sigma}\right) = \frac{1}{(2\pi)^{1/2}} \exp\left(\frac{-S^2}{2\sigma^2}\right) \tag{5.3}$$

By integrating and combining positive and negative values, the Gaussian exceedance function (Fig. 5.4b) is given by

$$P\left(\frac{S}{\sigma}\right) = \frac{2}{(2\pi)^{1/2}} \int_{S/\sigma}^{\infty} \exp\left(\frac{-S^2}{2\sigma^2}\right) d\left(\frac{S}{\sigma}\right) \tag{5.4}$$

Most random processes encountered are Gaussian, and this distribution applies to the instantaneous values of both broad-band and narrow-band random processes.

In fatigue, the distribution of peak values is usually of interest. For a narrow-band Gaussian process, the probability density function of the occurrence of a positive going peak of amplitude S_a approaches the Rayleigh distribution[10,21] (Fig. 5.4c)

$$p\left(\frac{S_a}{\sigma}\right) = \frac{S_a}{\sigma} \exp\left(\frac{-S^2_a}{2\sigma^2}\right) \tag{5.5}$$

as the band width decreases and the irregularity factor tends to 1.

The narrow-band random process is symmetrical about its mean value S_m from which S_a is measured, so corresponding negative peaks also appear, although this is not necessarily true for the general case of a Rayleigh distribution.[19] The corresponding exceedance function (Fig. 5.4d) is

$$P\left(\frac{S_a}{\sigma}\right) = \exp\left(\frac{-S_a^2}{2\sigma^2}\right) \tag{5.6}$$

If the resonant system is not sharply tuned, the peak amplitudes vary more rapidly and their distribution deviates[21] from equation (5.5). In the limit, as the band width tends to infinity and the irregularity factor to zero (white noise), the distribution of peaks tends to the Gaussian distribution given by equations (5.3) and (5.4). Both distributions theoretically extend to infinite stress, but for various practical reasons[1] peaks do not exceed a cut-off value of S_a/σ known as the clipping ratio, which is usually between 3 and 5.

5.3   SCATTER IN CONSTANT AMPLITUDE FATIGUE DATA

A great deal of effort has been devoted[1] to the determination of the distribution of the fatigue lives of nominally identical fatigue specimens tested under the same nominal constant amplitude loading. In many cases, it is found that the logarithm of the life distribution is a good approximation to the Gaussian or normal distribution. This log normal distribution may be written as

$$p(\log N) = \frac{1}{\sigma(2\pi)^{1/2}} \exp\left\{ -\tfrac{1}{2}\left(\frac{\log N - \overline{\log N}}{\sigma}\right)^2 \right\} \tag{5.7}$$

where $\overline{\log N}$ is the mean log life and σ the standard deviation, which provides a measure of the width of the scatter band. Large numbers of specimens are sometimes tested in attempts to find the precise form of the probability density function in given circumstances; some typical results[22] are shown in Fig. 5.5.

A life distribution function, supposedly giving better correlation with experimental data than the log normal distribution, has been proposed by Weibull[23]

$$p(N) = \frac{b}{N_a - N_0}\left(\frac{N - N_0}{N_a - N_0}\right)^{b-1} \exp\left\{ -\left(\frac{N - N_0}{N_a - N_0}\right)^b \right\} \tag{5.8}$$

where N_0, N_a and b are constants determined from the experimental data. The cumulative distribution function corresponding to the

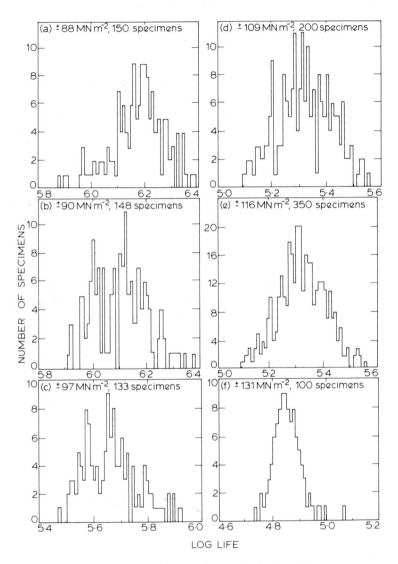

Fig. 5.5   Distribution of the fatigue lives of plain copper specimens at various stress levels (from Korbacher[22])

Weibull distribution function for the fraction of the batch of specimens which fail prior to N is

$$F(N) = 1 - P(N) = 1 - \exp\left\{ -\left(\frac{N - N_0}{N_a - N_0}\right)^b \right\} \qquad (5.9)$$

which allows a simple graphical method to fit data to the distribution and hence to estimate the unknown parameters b, N_0 and N_a. Because this relationship has three parameters, it might be expected to give a better fit to the experimental data than the log normal distribution which has only two (equation (5.7)). Even so, the complex nature of fatigue means that actual distributions must be complex (next section), so such relationships can only be regarded as approximations.[24] The assumption of a particular type of distribution can therefore lead to error, especially if predictions are made at the extreme limits of the distribution, i.e. at either very high or very low probabilities of failure.[25]

If groups of specimens of a particular material are tested at different stress levels, S/N curves for different probabilities of failure can be drawn. Such curves have been referred to as P–S–N curves.[23] Schematic examples are shown in Fig. 5.6 and some actual data[26] in Fig. 5.7. The curve for 50% probability of failure is the median, or middlemost life, and is what is aimed for when drawing a conventional S/N curve through a set of data.

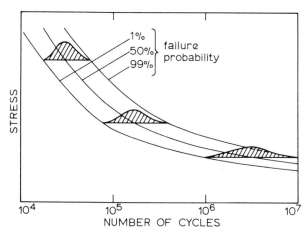

Fig. 5.6   Schematic P–S–N diagram showing the log normal distribution of lives at various stress levels (from Frost, Marsh and Pook[1])

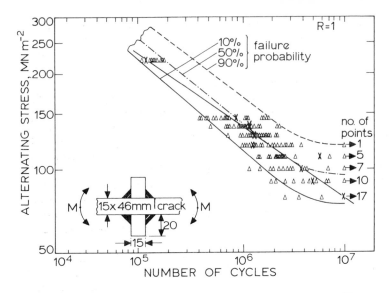

Fig. 5.7   Test results for welded joints (from Haibach[26])

Scatter in endurance increases as the fatigue limit is approached, as shown in Figs. 5.6 and 5.7. Statistical analysis in the vicinity of the fatigue limit is difficult.[9] This is because data are censored, i.e. some specimens are unbroken, so that it is only known that the life exceeds a certain number of cycles. A practical method is available.[27] There is greater scatter in the lives of a group of specimens tested at a stress level greater than their fatigue limit than in those tested at the stress levels necessary to cause failure in a given endurance. For example, although a difference of 10:1 in the lives of specimens tested at a given stress amplitude near the fatigue limit is not uncommon, it would be unusual for the ratio of the highest to the lowest fatigue limits obtained from different groups of specimens to be as high as $1\frac{1}{4}$.

Smooth S/N curves can usually be drawn without difficulty through a few test points. This is due[1] to the high expectancy that, with a small number of test results, a large proportion will fall fairly close to the most probable values. Figure 5.7 gives an example. One result was selected at each load level with the aid of a table of random numbers. These points are indicated with a cross and, with one exception, fall close to a smooth curve.

The basic principle of what is known as the analysis of variance[28] is that a group of one or more specimens is merely a sample taken from a large body or population. Such a sample is considered to be just one of a number of samples that could be tested. The results obtained from tests on a random sample from the population can be used to estimate the characteristics of the whole population, and to measure the reliability of the estimates. The values of the parameters for the population can only be estimated from tests on the sample; to obtain exact values would require the whole population to be tested. These estimates of the behaviour of the population from tests on a sample, and the confidence that can be placed in them, are the essence of the analysis of variance.[6,7,23,28,29] For example, calculation of the change in life that might be expected from small variations in nominal testing conditions easily shows[1] that all the scatter cannot be accounted for by expected differences inherent in the experimental techniques used in testing the specimens.

It is often necessary to know whether some procedure has resulted in a significant improvement in fatigue life at a given stress level. If the distribution is assumed to be normal (or log normal), standard statistical methods can be applied to determine whether observed differences can reasonably be ascribed to the scatter associated with fatigue, or if a worthwhile improvement has been achieved. For example, statistical tests have shown[30] that bevelling the ends of a particular type of glued lap joint is not worthwhile. Systematic differences often exist between tests carried out on the same batch of material at different laboratories, and on different batches of the same material.[24]

5.4 COMBINED DISTRIBUTIONS

The total life of a plain specimen comprises two stages: microcrack development and macrocrack propagation (Section 2.3). It is generally accepted[1] that most of the scatter in the lives of specimens tested at a given stress amplitude is associated with the development of a surface microcrack to the macrocrack stage. Thus, scatter is greater in plain specimens than in sharply notched specimens because a greater proportion of the total life is spent in developing a microcrack to the macrocrack stage.

Because the total life of a plain specimen comprises two stages, the distribution of the lives of plain specimens is an additive distribution made up from the distributions associated with the two stages which are statistically independent. Data such as those shown in Fig. 5.5 must therefore be interpreted in this light. At high stress levels, the number of cycles needed to initiate a crack is negligible, so that the S/N curve shows the relatively narrow scatter in life associated with crack propagation. At low stress levels, the number of cycles needed to propagate the crack is negligible, so that the S/N curve shows the wide scatter in life associated with crack initiation. At intermediate stresses, crack initiation and propagation are of roughly equal importance and the resultant additive distribution can show the bimodal (two-humped) form often associated with the presence of two different mechanisms. Thus, the data may be interpreted schematically as shown in Fig. 5.8; however, it is not possible to separate an additive distribution into its components without prior knowledge of the components.[10]

In testing structures, it is often found that one fracture mode predominates at high stresses and another at low stresses. For instance, in the testing of riveted sheet structures, rivet failure is associated with high stresses and sheet failure with low stresses.[1] This is shown schematically in Fig. 5.9 where the underlying distribu- ʻions are assumed to be log normal. At stress levels where the scatter bands do not overlap, failure is entirely by the mode corresponding to the shorter life. In the vicinity of the cross-over point the scatter bands overlap, and a proportion of the specimens will survive and

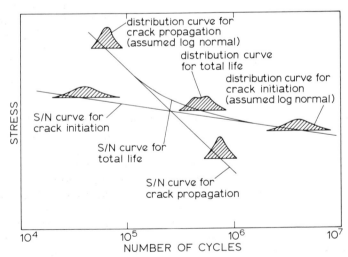

Fig. 5.8   Schematic diagram showing the *S/N* diagram for a combination of *S/N* curves for crack initiation and crack propagation (from Frost, Marsh and Pook[1])

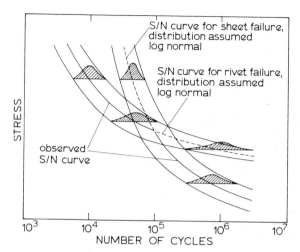

Fig. 5.9 Schematic diagram illustrating the behaviour of riveted sheet structures (from Frost, Marsh and Pook[1])

fail by the mode corresponding to longer life. As only a proportion survive to fail by the longer life mode, the two distributions corresponding to the two modes are not statistically independent, and the longer life mode is a conditional distribution.[10] The resulting combined distribution is therefore a skewed (Fig. 5.9) rather than the bimodal distribution associated with the combination of two statistically independent processes. The overall S/N curve has a discontinuity at the cross-over point.

5.5 SCATTER IN FATIGUE CRACK GROWTH DATA

From a statistical viewpoint, a fatigue crack growth test has important differences from a traditional fatigue test, so the established methods of applying statistics to fatigue (Sections 5.1–5.4) are not appropriate. In constant amplitude fatigue crack growth rate testing,[1,16] each experiment aims to establish the relationship between crack length, a, and number of cycles N. Measurements of crack length and number of cycles are made at intervals which are usually small enough for ΔK_I to be essentially constant during an interval. Values of ΔK_I and da/dN are obtained for various crack lengths, and data from one or more specimens are used to determine the relationship between ΔK_I and da/dN plus minor variables (Section 4.5). Values of da/dN are obtained either by direct calculation between successive pairs of readings or by fitting a curve to the basic a versus N data and taking tangents. The set of points obtained may then be subject to a further curve-fitting technique, usually with the assumption that the data follow a particular law such as equation (4.4). Subjective judgments are involved and a completely satisfactory procedure has yet to be developed.

Fatigue crack growth data obtained by different investigators on similar materials usually agree reasonably well, although there may be considerable variation in the amount of scatter. Without some knowledge of the statistics of individual sources of scatter it is not possible to derive information on them from an examination of the observed crack growth data. However, it has been observed[31] that the amount of scatter in fatigue crack growth data depends on material purity, and the amount of scatter observed in fatigue crack growth experiments is generally considerably greater than can be accounted for by experimental inaccuracies. It is reasonable to conclude, as has been previously suggested,[12,32] that the remaining scatter is due to the essentially random nature of fatigue crack growth, which is a consequence of the irregular structure of metals. Any attempt to analyse scatter must clearly take this into account. In understanding the sources of scatter in fatigue crack growth rate data, the first requirement is an appropriate statistical description of the shape and position of the crack front at the end of each cycle. It should be precise enough to be useful, yet loose enough to recognize practical limitations in the characterization of fatigue crack growth.

5.5.1 Description and measurement of cracks

Crack growth can be viewed at a range of scales[33] (Fig. 4.1). On a macroscopic scale (10^{-1} mm), a metal can normally be regarded as a homogeneous continuum, and the size and shape of a crack are usually intuitively obvious, with there being no real difficulty in measuring the position of a crack tip at a given time. On a microscopic scale, cracks are very irregular. In consequence, there can be a real difficulty in defining what is meant by crack length or area. The problem is analogous to the question of how long a coastline is (see Fig. 4.2). The quoted length of a crack, or coastline, has no precise meaning unless the method of measurement and a characteristic length are specified.[14] The result obtained increases as the characteristic length is decreased, i.e. the shape is examined in more detail, perhaps by the use of increased magnification. A simple approach is to approximate the line by straight steps of specified size with ends lying on the line, and to take the length as the sum of the step lengths. For example,[14] for step lengths η of 1–1 000 km, the length $L(\eta)$ of the west coast of Britain is proportional to η^{1-D}, where D is the fractal dimension, which in this case is equal to about $1\frac{1}{4}$. Such irregular lines are examples of fractal objects. The characterization of fracture surfaces using the theory of fractals has been discussed by Chermant and Coster,[13] who found that D was about 1·1 for sections through intergranular fractures. The statistical descriptions used are related to Fourier analysis[14] and can, in principle, be extended to include descriptions of sections through branched cracks, and to crack surfaces. As so far developed, the concept of the fractal object is too general to be helpful in the characterization of fatigue fracture surfaces, where the position of the crack front at the end of successive cycles needs to be considered, but it does help to clarify the nature of the problem.

Many different methods can be used to monitor crack length during a fatigue test.[34] Two common methods are optical observation of the specimen surfaces and the potential drop technique. In the latter technique, a high constant current is applied through leads at the specimen ends; potential leads are located at each side of the crack to measure the change in potential drop across the crack as it grows. Crack lengths can be measured to accuracies of around 10^{-2} mm, although the potential drop technique[35] can be used to resolve increments of crack growth due to individual cycles at crack growth rates down to 10^{-4} mm/cycle. Fractographic examination[36] appears capable of resolution of the striations found on fatigue fracture surfaces at crack growth rates down to the order of 10^{-3} mm/cycle.

Successive positions of the crack front, projected onto the macroscopic plane of the crack, can sometimes be obtained fractographically, but a great deal of effort is involved. Figure 5.10 shows an example, derived from Fig. 2.4 of Ref. 37. Successive positions of the crack front form an ensemble of random processes which may be characterized by the standard methods[10,11] often used in random vibration and electronics applications.

The randomness of individual crack fronts may be described by the auto-correlation function, and the relationships between neighbouring crack fronts by the cross-correlation function. Scatter in fatigue crack growth data is usually assumed to be log normal,[1] so, defining δa as the mean spacing between adjacent positions of the crack front, it is appropriate to assume that the distribution of δa is log normal. This recognizes that da/dN cannot be negative. Although the necessary statistical parameters cannot, in practice, be deter-

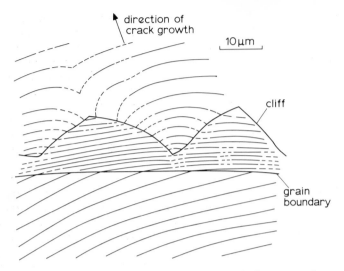

**Fig. 5.10   Successive positions of the crack front on the
fracture surface of an Al–7·5%Zn–2·5%Mg
corrosion fatigue specimen**

mined, it is possible to draw useful qualitative conclusions on
scatter.[12]

5.5.2   Analysis of scatter

Consider an idealized fatigue crack growing across a plate under
constant amplitude loading. It is assumed that on a macroscopic
scale it is flat and perpendicular to the applied stress, with the crack
front straight and perpendicular to the direction of crack growth.
Using the small-scale argument (Section 4.3), ΔK_I can be considered
to be constant along the crack front. It is also assumed that the crack
is long so that $d(\Delta K_I)/da$ is small, and ΔK_I can be taken as constant
over small increments of crack growth. As history effects can be
important (Chapter 6), it must also be assumed that the crack has
been propagating long enough for a steady state to have been reached.
These assumptions ensure that the fatigue crack growth can be
regarded as statistically stationary, i.e. the statistical parameters do
not change as the crack grows.[12] In this ideal case there are no
difficulties in defining macroscopically what is meant by crack length.
    There is usually considerably more scatter in data obtained on
thin sheets using optical crack length measurement (e.g. Ref. 38)

than in data obtained on thick specimens using the potential drop technique (e.g. Ref. 35). While some of this difference can no doubt be accounted for by experimental inaccuracies, some of it is due to differences in experimental technique.[12] The potential drop technique can be arranged[35] so that an average is taken over the whole crack; apart from experimental error, the average crack length is therefore measured precisely. Optical measurement only samples points on one or both surfaces, thus only an estimate of the average crack length is obtained, and the apparent scatter increases.[12] Systematic errors are introduced if the crack front changes shape; apparent scatter at the start of a fatigue crack growth test is often caused by changes of crack front shape.[1,39]

It is, in principle, easy to obtain statistics of the variability of fatigue crack growth rates by measuring the crack length at appropriate intervals. Unfortunately, it follows from random process theory[40] that measurements should be made at least once per cycle with an accuracy much greater than is at present possible (see previous section), making this impractical. However, it can be shown[12] that the scatter due to the random nature of fatigue crack growth is inversely proportional to the square root of the number of cycles (or crack length) over which an average value of da/dN is calculated.

The aim of curve fitting is ostensibly to reduce scatter in data by reducing the effect of experimental variations and measurement errors; however, considerable difficulty is encountered[41,42] in defining the criterion to be adopted in assessing the acceptability of a curve-fitting procedure, which may be nothing more than drawing a smooth curve by eye. Curve-fitting techniques increase the effective number of cycles over which an average value of da/dN is calculated. It is this, as well as the minimization of the effect of experimental inaccuracies, which accounts for the apparent reduction in scatter as more carefully fitted curves are used,[12] and shows why methods of reducing fatigue crack growth data cannot avoid making use of subjective judgments.

The dependence of scatter in da/dN versus ΔK_I data on precise data reduction techniques can lead to problems in the estimation of values of m and D in equation (4.4), or in fitting more elaborate equations to various parts of the da/dN versus ΔK_I relationship. How this can arise[16] is shown schematically in Fig. 5.11. Theoretically (Section 4.9), a slope corresponding to $m = 2$ would be expected with cut-offs at the upper and lower ends corresponding to static failure and the threshold, respectively. In practice (Section 4.5.1), these will normally be approached gradually, as indicated by the heavy line,

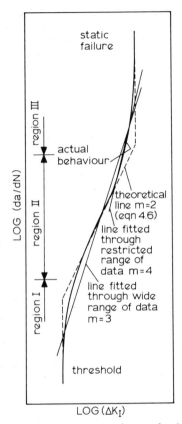

Fig. 5.11 Relationship between theoretical and experimental values of _m_ (from Chell[16])

giving an overall sigmoidal curve. Scatter usually prevents determination of the precise shape of the curve, and fitting straight lines to different parts of the data will often give different values for _m_ and _D_, as indicated in Fig. 5.11. For this reason, it is perhaps preferable to quote data, as in Table 4.1, for the value of _m_ together with the value of ΔK_I at a convenient rate of crack growth. Fitting a straight line to a wide range of data usually gives a value for _m_ of around 3.

More elaborate equations which take into account curvature at either end of the data are sometimes used. The utility of such equations, all of which are empirically based, must be judged by the intended application, the most important point being to ensure that they are not extrapolated beyond the range of the experimental data.[43] A convenient, easily fitted model based on adding the material's resistance to fatigue crack growth (i.e. $(da/dN)^{-1}$) in the three regions shown in Fig. 5.11 has been developed.[44] The characteristic equation describing da/dN for a constant load ratio R and fixed environment is

$$\frac{1}{(da/dN)} = \frac{A_1}{(\Delta K_I)^{n_1}} + \frac{A_2}{(\Delta K_I)^{n_2}} - \frac{A_2}{\{K_{c,f}(1-R)\}^{n_2}} \qquad (5.10)$$

where A_1, n_1, A_2, n_2 and $K_{c,f}$ are constants that can be obtained from the data. The three terms correspond to the three regions of crack growth; transition regions are modelled by a combination of two adjacent terms. The exponents n_1 and n_2 can be estimated from the slopes of plotted data. The value of $K_{c,f}$, the fatigue toughness, which should not be confused with the fracture toughness, can also be determined from plotted data. If growth rates are not to be characterized in region III (Fig. 5.11), the last term is set to zero. If A_1 is also set to zero, the equation reduces to equation (4.4). The threshold is not adequately modelled in that a positive value of da/dN is predicted however low the value of ΔK_I. This is claimed[44] to have the advantage that extrapolation to low ΔK_I values is conservative.

5.6 SCATTER IN FATIGUE CRACK GROWTH LIFE

The large amount of scatter which can be present in fatigue crack growth data is illustrated by Fig. 4.9 which shows test results for mild steel. The boundaries of the scatter band were drawn to include approximately 90% of the data. However, it is generally accepted[1] that the crack growth phase in a fatigue test on a plain specimen contributes much less to overall scatter than the crack initiation phase (Section 5.4). In addition, it has been observed[45] that a factor of 2 on fatigue crack growth rates is an adequate allowance for scatter in estimating inspection intervals for aircraft.

The apparent paradox is resolved when data such as those given in Fig. 5.12 are examined.[12] Crack growth curves for duplicate tests on 18/8 austenitic steel are shown. An examination of adjacent points shows that the slopes of the two curves can vary widely at equivalent points, yet there is very little difference in total life, which illustrates why there is no simple relationship between the scatter observed in fatigue crack growth rate data and the scatter in total life.

In practical engineering problems (Section 4.6), the amount of scatter in the number of cycles required for a crack to grow from

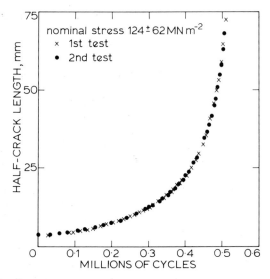

Fig. 5.12   Crack growth curve for two similar 18/8 austenitic steel sheets (from Pook[12])

some initial length a_0 to a much greater final length a_f is usually of interest. Sometimes, the initial[46] and final[47] lengths will be known only as distributions. Assuming that a_0 and a_f are known exactly, the mean total life N_f is given by the generalized form of equation (4.5)

$$N_f = \int_{a_0}^{a_f} \overline{(dN/da)}\, da \qquad (5.11)$$

where $\overline{dN/da}$ is the mean value of dN/da at a particular crack length.

The distribution of N_f about its mean value $\overline{N_f}$ is clearly of interest, in particular the coefficient of variation (standard deviation/$\overline{N_f}$); it can be examined by considering the statistics of increments of crack growth in equation (5.11). The distribution of N_f is then obtained as an additive distribution made up of the distributions for all the increments. Lack of information prevents evaluation of these; however, as the coefficient of variation for individual increments[12] is proportional to $n^{-1/2}$, where n is the number of cycles in the increment, it can be seen that the overall coefficient of variation decreases as $(a_f - a_0)$ increases, which is in accordance with practical observations. Some guidance is provided by the observation that scatter bands for fatigue crack growth data are generally parallel sided when plotted on logarithmic coordinates,[1,38] e.g. Fig. 4.9, which suggests that the coefficient of variation of dN/da is independent of dN/da. This in turn suggests that the coefficient of variation of N_f will be approximately inversely proportional to $(a_f - a_0)^{1/2}$. In fact, by judicious selection of the increments of crack length over which da/dN in fatigue crack growth tests is averaged, empirical correlations between the scatter in crack growth data and total life can be obtained.[48]

In practice, scatter bands for fatigue crack growth data often incorporate data obtained from different batches of nominally similar materials, and part of the scatter can be attributed to batch-to-batch variations in material behaviour. It is not in general possible to separate the scatter due to material variation from that due to the random nature of fatigue crack growth. Variations in fatigue crack growth rates appear directly as variations in calculated lives. The implicit, and conservative, assumption is often made that scatter is due to variations in material behaviour, which is equivalent to the incorrect assumption that fatigue crack growth behaviour is not random. This conservative assumption is commonly employed in safety assessments by using the upper bound of the scatter band in calculations.[49] More sophisticated, though not necessarily more reliable, calculations make use of assumed distributions for fatigue crack growth data.[50]

REFERENCES

1   N. E. FROST, K. J. MARSH and L. P. POOK: 'Metal fatigue', 1974, Oxford, Clarendon Press
2   BRITISH STANDARDS INSTITUTION: 'Methods of fatigue testing. Part 5: Guide to the application of statistics', BS 3518: Part 5: 1966
3   'Guide to fatigue testing and statistical analysis of fatigue data', ASTM STP 91-A, 2nd ed., 1963, Philadelphia, Pa, American Society for Testing and Materials
4   T. R. GURNEY: 'Fatigue of welded structures', 2nd ed., 1979, London, Cambridge University Press
5   BRITISH STANDARDS INSTITUTION: 'Steel, concrete and composite bridges. Part 10: Code of practice for fatigue', BS 5400: Part 10: 1980
6   L. G. JOHNSON: 'The statistical treatment of fatigue experiments', 1964, London, Elsevier
7   R. E. LITTLE and E. H. JEBE: 'Statistical design of fatigue experiments', 1975, London, Applied Science Publishers Ltd

8   M. G. KENDALL and A. STUART: 'The advanced theory of statistics', 1972, London, Griffin and Company

9   J. E. SPINDEL and E. HAIBACH: *Int. J. Fatigue*, 1979, **1**, (2), 81–88

10 A. PAPOULIS: 'Probability, randon variables and stochastic processes', 1965, London, McGraw-Hill

11 J. S. BENDAT and A. G. PIERSOL: 'Random data: analysis and measurement procedures', 1971, New York, Wiley–Interscience

12 L. P. POOK: *J. Soc. Environ. Eng.*, 1976, **15-4**, (71), 3–10

13 J. L. CHERMANT and R. COSTER: *J. Mater. Sci.*, 1979, **14**, (3), 509–534

14 B. B. MANDELBROT: 'Fractals: form, chance and dimension', 1977, San Francisco, W. H. Freeman

15 S. P. TIMOSHENKO and J. N. GOODIER: 'Theory of elasticity', 3rd ed., 1970, New York, McGraw-Hill

16 G. G. CHELL (ed.): 'Developments in fracture mechanics—I', 1979, London, Applied Science Publishers, Ltd

17 S. H. SMITH: 'Acoustical fatigue in aerospace structures' (eds. J. W. Trapp and P. M. Forney), 331–360, 1965, New York, Syracuse University Press

18 L. P. POOK: *J. Soc. Environ. Eng.*, 1978, **17-1**, (76), 22–23, 25–28, 31–35

19 F. SHERRAT and P. R. EDWARDS: *ibid.*, 1974, **13-4**, (63), 3–14

20 W. D. DOVER and N. F. BOUTLE: *J. Strain Anal.*, 1978 **13**, (3), 129–139

21 D. E. CARTWRIGHT and M. S. LONGUET-HIGGINS: *Proc. R. Soc.*, Series A, 1956, **237**, 212–232

22 G. K. KORBACHER: *Exp. Mech.*, 1971, **11**, (12), 540–547

23 W. WEIBULL: 'Fatigue testing and analysis of results', 1961, Oxford, Pergamon Press

24 R. E. LITTLE: *ASTM Standardization News*, 1980, **8**, (2), 23–25

25 N. T. BLOOMER and T. F. ROYLANCE: *Aeronaut. Q.*, 1965, **16**, (4), 307–322

26 E. HAIBACH: Proc. of Conf. on Fatigue of Welded Structures, Vol. II, 35–39, 1970; 1971, Cambridge, The Welding Institute

27 R. E. LITTLE: *J. Test. Eval.*, 1980, **8**, (2), 80–84

28 C. LIPSON and N. H. SHETH: 'Statistical design and analysis of engineering experiments', 1973, New York, McGraw-Hill

29 S. R. SWANSON (ed.): 'Handbook of fatigue testing', ASTM STP 566, 1974, Philadelphia, Pa., American Society for Testing and Materials

30 D. R. HAMEL, G. K. KORBACHER and D. M. SMITH: *J. Bas. Eng.*, 1971, **93D**, (4), 649–656

31 P. R. V. EVANS, M. B. OWEN and B. E. HOPKINS: *Eng. Fract. Mech.*, 1971, **3**, (4), 463–473

32 L. P. POOK and N. E. FROST: *Int. J. Fract.*, 1973, **9**, (1), 53–61

33 F. A. McCLINTOCK and G. R. IRWIN: 'Fracture toughness testing', ASTM STP 381, 84–113, 1967, Philadelphia, Pa., American Society for Testing and Materials

34 C. J. BEEVERS (ed.): 'The measurement of crack length and shape during fracture and fatigue', 1980, Warley, West Midlands, Engineering Materials Advisory Services Ltd,

35 P. McINTYRE: *J. Soc. Environ. Eng.*, 1974, **13-3**, (62), 3–7

36 R. C. BATES, W. G. CLARK and D. M. MOON: 'Electron microfractography', ASTM STP 453, 192–214, 1969, Philadelphia, Pa., American Society for Testing and Materials

37 C. A. STUBBINGTON: Report No. CPM4, 1963, Farnborough, Royal Aircraft Establishment

38 N. E. FROST, L. P. POOK and K. DENTON: *Eng. Fract. Mech.*, 1971, **3**, (2), 109–126

39 N. E. FROST: First Int. Conf. on Fracture, Sendai, Japan, 1965, **3**, 1433–1459

40 L. P. POOK: 'Fatigue-crack-growth measurement and data analysis' (eds. S. J. Hudak and R. J. Bucci), ASTM STP 738, 203–204, 1981, Philadelphia, Pa., American Society for Testing and Materials

41 W. G. CLARK and S. J. HUDAK: *J. Test. Eval.*, 1975, **3**, (6), 454–476

42 K. B. DAVIES and C. E. FEDDERSEN: *Int. J. Fract.*, 1973, **9**, (1), 116–118

43 D. W. HOEPNER and W. E. KRUPP: *Eng. Fract. Mech.*, 1974, **6**, (1), 47–70

44 A. SAXENA and S. J. HUDAK: *J. Test. Eval.*, 1980, **8**, (3), 113–118

45 A. J. TROUGHTON and J. McSTAY: 'Current aeronautical fatigue problems', 429, 1965, Oxford, Pergamon Press

46 W. H. TANG: *J. Test. Eval.*, 1973, **1**, (6), 459–467

47 T. R. BRUSSAT: 'Damage tolerance in aircraft structures', ASTM STP 486, 122–147, 1971, Philadelphia, Pa., American Society for Testing and Materials

48 W. J. D. SHAW and I. LE MAY: *Fatigue Eng. Mater. Struct.*, 1981, **4**, (4), 367–375

49 P. C. RICCARDELLA and T. R. MAGER: 'Stress analysis and growth of cracks', ASTM STP 513, 260–279, 1972, Philadelphia, Pa., American Society for Testing and Materials

50 R. A. SMITH (ed.): 'Fracture mechanics: current status, future prospects', 1979, Oxford, Pergamon Press

CHAPTER 6

Fatigue crack growth under non-uniform conditions

Variety's the very spice of life,
That gives it all its flavour.
W. Cowper, *The Task*.

6.1 INTRODUCTION

Two plasticity effects can have a significant effect on fatigue crack growth behaviour and must be considered in the discussion of fatigue crack growth under non-uniform conditions. One is caused by the residual compressive stresses due to reverse yielding at the crack tip after unloading (Section 4.4); for an initially stress-free crack, these mean that a compressive load is required to completely close the crack. The other is that a propagating fatigue crack leaves behind a wake of plastically deformed material adjacent to the fracture surfaces. This wake is responsible for the phenomenon of crack closure, mentioned in Section 4.5.2, which has been extensively discussed in the literature since it was described by Elber.[1] It was apparently first described by Frost,[2] although not under that name. This wake may cause the crack to close at above the minimum load in the fatigue cycle, thus reducing the effective value of ΔK_I (ΔK_{eff}) to below the conventionally calculated value (Section 4.5.2). In practice, both crack closure[3] and deformations near a crack tip[4] may be complex (Figs. 6.1 and 6.2), making a detailed analytic approach impossible. However, the main effects are conveniently illustrated by a set of elastic–plastic finite element calculations;[5] various parameters were adjusted to model effects observed in practice, except that the numbers of cycles used were less than are actually involved.

In the discussion of fatigue crack growth under constant amplitude loading it is normally assumed (Sections 4.5.2 and 5.5.2) that the value of ΔK_I does not change significantly over short distances, i.e. $d(\Delta K_I)/da$ is small. It is also assumed that the crack has been growing long enough for the plastic wake to have reached a stable condition. The crack may therefore be taken as growing under steady-state

98

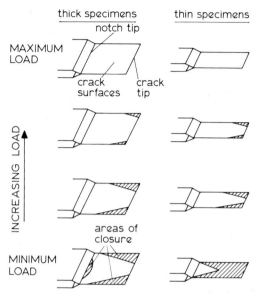

Fig. 6.1   Crack closure behaviour of aluminium alloy compact tension specimens, $R = 0\cdot01$ (based on Lafarie-Frenot, Petit and Gasc[3])

conditions, i.e. it is a statistically stationary process (Section 5.2). Anything that disturbs this steady state disturbs the plastic zone and its wake, and is likely to affect the fatigue crack growth rate. In this chapter, ΔK_I is the conventionally calculated value (Section 4.5.2) which neglects plasticity effects.

6.2   CRACK GROWTH FROM AN INITIALLY STRESS-FREE CRACK

For the case of a minimum load in the fatigue cycle of zero ($R = 0$), the simple model of fatigue crack growth described in Section 4.9 makes no distinction between steady-state crack growth and growth from a crack initially free from residual stresses and plastic deformation. In practice, the development of a plastic wake means that crack closure occurs at progressively higher stresses, with concomitant reductions in ΔK_{eff} and hence the rate of crack growth, until a steady state is reached, as illustrated in Fig. 6.3. For most constant amplitude loading situations it is possible to neglect the transient region before steady-state conditions are reached.

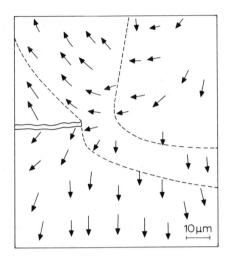

Fig. 6.2   Crack tip displacements in an aluminium alloy due to changing K_I from 3·8 to 13·8 MN m$^{-3/2}$ (actual values × 10) (from Davidson and Lankford[4])

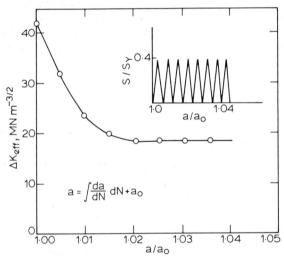

Fig. 6.3   Variation of ΔK_{eff} with cyclic crack growth for constant amplitude loading (from Nakagaki and Atluri[5])

In terms of the conventionally calculated value of ΔK_I, it is clear that an R-curve effect, in which resistance to crack growth increases with crack length, exists for fatigue crack growth.[6] The R-curve concept as originally proposed[7] suggests that resistance to crack growth under a monotonically increasing static load increases with crack length, and is a function of the amount of crack growth δa, rather than the initial crack length a_0. In certain circumstances, a fatigue crack may stop growing, leading to a 'non-propagating' fatigue crack.

A type of crack closure argument, based on the difference between steady-state crack growth and crack growth from an initially stress-free crack, can be used in conjunction with the model described in Section 4.9 to explain[8] why fatigue crack growth rates are often largely independent of mean stress.

Threshold behaviour represented by the R-curve concept[9] is shown schematically in Fig. 6.4. The R-curve gives the minimum value of ΔK_I required to cause further crack growth after an increment of crack growth, δa. The shape of the R-curve is assumed to be independent of the initial crack length a_0, and the specimen is assumed to be initially free from residual stresses in the vicinity of the crack tip, such as those resulting from previous fatigue loading. The condition for crack growth on the first cycle corresponds to the need for an increment of crack growth to be at least one lattice spacing (Section 2.3). By analogy with the original proposal[7] for static loading, the R-curve could in principle be determined for a fatigue

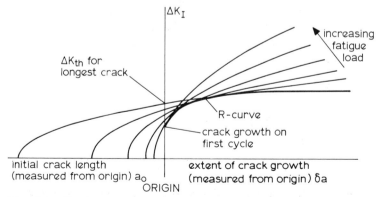

Fig. 6.4 R-curve concept applied to the fatigue crack growth threshold (from Pook and Greenan[9])

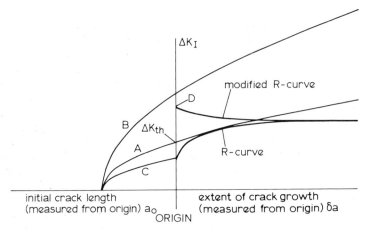

Fig. 6.5   Modified R-curve for prior loading (from Pook and Greenan[9])

loading whose severity was increased as the crack grew such that it was just enough to cause continued crack growth. For the fatigue crack growth threshold it is more practical to define the R-curve as the envelope of a family of ΔK_I against crack length curves which are just sufficient to cause continued crack growth. In a constant amplitude test on a stress-relieved cracked specimen (Section 4.7), the threshold condition is reached when a curve of ΔK_I against crack length is tangential to the R-curve (curve A, Fig. 6.5). It is assumed that the curve eventually levels off to an upper shelf value and that S_m/S_a is constant. If the ΔK_I curve were higher (curve B, Fig. 6.5) the specimen would fail, and if it were lower (curve C) the specimen would remain intact, although there might be a limited amount of crack growth producing a non-propagating fatigue crack. The variation of ΔK_{th} with S_m/S_a (Section 4.7) can be modelled as a variation in the general level of the R-curve. Noticeable non-propagating cracks that occur in certain corrosive environments[9] can be modelled as a change in the shape of the R-curve. Except for short crack lengths (Section 6.4), and in corrosive environments, the upper shelf of the R-curve can be taken as a geometry-independent value of ΔK_{th} for most practical applications.

6.3   EFFECT OF LOAD CHANGES

In general,[10] a crack will continue to grow at the expected rate when the load level is increased, but will be retarded when the load level is reduced. Prior loading at a higher level increases the fatigue crack growth threshold. Following an increase in load level, there is a small transient region before the steady state corresponding to the new load level is reached. In this transient region, which can often be neglected, the amount of crack closure is reduced (Fig. 6.6) and the crack growth rate correspondingly increased to above the expected rate. Following a reduction in load level, the residual compressive stresses ahead of the crack tip due to the prior loading substantially increase the amount of crack closure (Fig. 6.7), which reduces the crack growth rate to well below that expected. The effect persists until the plastic zone associated with the new load level reaches the edge of that associated with the prior loading. A single overload has a similar effect (Fig. 6.8).

Numerous overload tests have been carried out but, in detail, behaviour is too complicated for the results of such comparatively simple tests to be used to predict fatigue crack growth under more complicated load histories. Data are usually presented as crack length versus number of cycles curves. Some typical results[11] are shown in Fig. 6.9. The baseline loading was a constant ΔK_I of 30 MN m$^{-3/2}$ with $R = 0.05$. Results are sometimes quoted as delay cycles, i.e. the

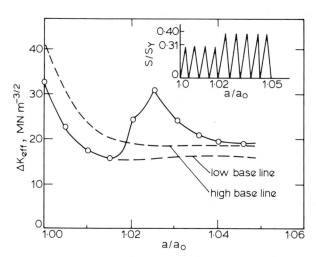

Fig. 6.6 Variation of ΔK_{eff} with cyclic crack growth for low-to-high block loading (from Nakagaki and Atluri[5])

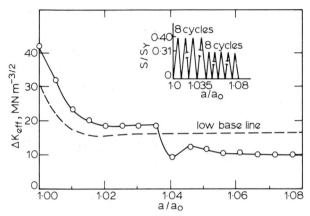

Fig. 6.7 Variation of ΔK_{eff} with cyclic crack growth for high-to-low block loading (from Nakagaki and Atluri[5])

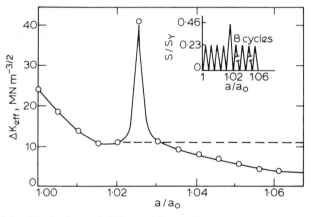

Fig. 6.8 Variation of ΔK_{eff} with cyclic crack growth for a single overload (from Nakagaki and Atluri[5])

additional number of cycles required for crack growth through the region affected by the overload.

The effect of prior loading on the threshold can be represented by a modified R-curve (Fig. 6.5); a curve of ΔK_I against crack length would become tangential at a point D giving the value of ΔK_{th} after

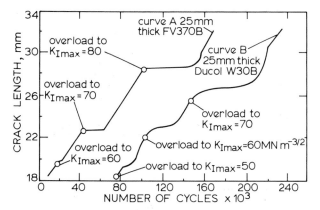

Fig. 6.9 **Typical crack growth retardation curves (from Bathias and Baïlon[11])**

prior loading. This shows why great care is needed in measuring the threshold by following the da/dN versus ΔK_I curve downwards (Section 4.7). If a K-decreasing test is used for this purpose $d(\Delta K_I)/da$ is negative, and it has been suggested[12] that to avoid obtaining spurious high values for the threshold, the expression

$$\left| \frac{1}{\Delta K_I} \frac{d(\Delta K_I)}{da} \right| \leqslant 0.08 \text{ mm}^{-1} \tag{6.1}$$

should be satisfied. This corresponds to an exponential decrease in ΔK_I with the value decreasing by half in not less than 8·7 mm.

6.4 THE SHORT-CRACK PROBLEM

Most fatigue crack growth data are obtained from specimens containing relatively long cracks. Usually, tests are carried out under constant amplitude loading and ΔK_I increases with crack length. Over small increments of growth, it is generally possible (Sections 4.5.2 and 5.5.2) to regard ΔK_I as constant, i.e. $d(\Delta K_I)/da$ can be regarded as small.

It has been known for some time that, for relatively short (less than roughly 1 mm) macrocracks, fatigue crack growth rates can be increased[13] and the threshold depressed[14] compared with the rates for long cracks. This can be due to the failure of stress intensity factors to provide an adequate representation of the crack tip stress field (Section 4.8), the coalescence of adjacent cracks (Section 4.9), or the difficulty of calculating accurate stress intensity factors for short cracks, which are often very irregular.[15] In welded joints, it may be due to the presence of residual stresses.[16] This section is concerned with short-crack effects that arise because $d(\Delta K_I)/da$ cannot be regarded as small. Threshold behaviour is discussed first, using the R-curve concept (Section 6.2).

For a short crack, ΔK_I increases rapidly with crack length and a curve of ΔK_I against crack length may become tangential to the R-curve before the upper shelf is reached (Fig. 6.4). The value of ΔK_{th} calculated from the initial crack length in tests based on the fatigue limit of precracked specimens (Section 4.7) will then be below the long crack value. Short, but easily detectable, non-propagating cracks are often found in sharply notched specimens which are unbroken after fatigue testing.[6] The stress intensity factor for a crack at a notch increases monotonically with crack length,[17] so non-propagating cracks cannot form at sharp notches if ΔK_{th} is independent of crack length.[6] The occurrence of noticeable non-propagating cracks in sharply notched specimens can be regarded as due to the particular shape of the ΔK_I against length curve, compared with that of the R-curve. The occurrence of non-propagating cracks could equally well be modelled by assuming that ΔK_{th} actually remains constant, but that the development of the plastic zone wake means that ΔK_{eff} decreases with crack length despite the increase in the nominal value of ΔK_I. Other parameters could also be used; what matters is that a crack will stop if a parameter representing the crack tip stress field drops below a critical value necessary for crack growth.

The high rates of crack growth associated with short cracks are readily explained. An increase in ΔK_I produces a temporary increase in ΔK_{eff} to above the corresponding steady-state value (Fig. 6.4). Therefore, if $d(\Delta K_I)/da$ is sufficiently great, the value of ΔK_{eff} always exceeds the steady-state value corresponding to the current value of ΔK_I. Differentiating equation (4.2), assuming α to be constant, gives

$$\frac{1}{\Delta K_I} \frac{d(\Delta K_I)}{da} = \frac{1}{2a} \qquad (6.2)$$

Substituting $a = 1$ mm leads, as a first approximation, to the conclusion that $d(\Delta K_I)/da$ cannot be regarded as small if

$$\frac{1}{\Delta K_I} \frac{d(\Delta K_I)}{da} \geq 0 \cdot 5 \text{ mm}^{-1} \qquad (6.3)$$

6.5 CRACK GROWTH UNDER VARIABLE AMPLITUDE LOADING

Most structures in service are subjected to varying amplitude loads and this complicates predictions. Variable amplitude loadings can be divided into two broad classes: those in which individual load cycles may be distinguished, such as narrow-band random loading (Fig. 5.3), and those in which individual cycles cannot be distinguished, such as broad-band random loading (Fig. 5.2). The methods of random process theory (Section 5.2.1) are needed for the precise characterization of such load histories. If realistic load histories are used to obtain fatigue crack growth rate data, and these are analysed using a convenient parameter such as the root mean square (rms) value of $K_I(K_\sigma)$, then methods of life prediction are essentially the same as for constant amplitude loading (Sections 4.6 and 5.6), and interaction effects between different load levels (Section 6.3) are automatically taken into account. Large numbers of variable amplitude fatigue crack growth tests (e.g. Ref. 18) have been carried out using test methods essentially the same as those used for constant amplitude loading (Section 4.5). If a programmed load history is used, macroscopic markings on the fracture surface corresponding to different parts of the programme (e.g. Fig. 2.3) can be used to derive fatigue crack growth rate data.[16] When comparing data, care must be taken to check the precise convention used in calculating K_σ, and for broad-band random loadings the convention used for calculating equivalent cycles.[19] Figure 6.10 shows some results[16] for a medium strength structural-steel weldment tested using the load history shown in Fig. 3.5. The mean load was zero and K_σ was calculated from the whole process. Crack sizes were obtained from programme markings (Fig. 2.3).

Under complicated load histories, interaction effects between different load levels can have a profound influence on crack growth behaviour, but despite extensive experimental work it is difficult to extract useful generalizations.[19] In particular, different load histories may rate materials in a different order of merit, leading to difficulties in material selection.[20] In general, it is not possible to predict variable amplitude fatigue crack growth behaviour from the results of constant amplitude tests, although various empirical correlations are possible, and some theoretical progress is being made.[21] Particular difficulties arise in thin sheets owing to the influence of the transition to slant crack growth.[22]

Where individual cycles may be distinguished, the simplest method of prediction is to assume that each cycle causes the same amount

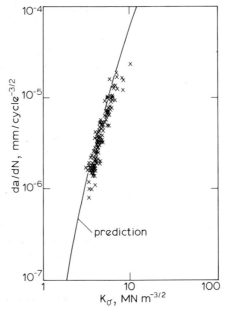

Fig. 6.10   Crack growth data for medium strength C–Mn structural steel under non-stationary narrow-band random loading (from Pook[16])

of growth as it would if applied as part of a sequence of loads of constant amplitude. This is analogous to a Miner's rule summation (Section 2.4). Under narrow-band random loading a cycle does not differ greatly from its predecessor (Section 5.2.1), and this simple approach can give reasonable results for materials such as mild steel where interaction effects are relatively unimportant.[23,24] The prediction shown in Fig. 6.10 was made on this basis. (It should be noted that such predictions are only mathematically correct[25] when crack growth data are represented by the Paris law (equation 4.4).) More generally, predictions are excessively conservative,[26] and interaction effects (Section 6.3) must be taken into account. Unfortunately, calculations can become very complicated.[20,27,28] Even for mild steel, interaction effects become important at very long ($\sim 10^8$ cycles) lives.[27]

A number of empirical methods of predicting fatigue crack growth behaviour under variable amplitude loadings have been developed.

All contain constants which are adjusted to suit the results of particular tests and permit extrapolation to broadly similar situations. The best known is due to Wheeler.[29] Each cycle is considered separately, and the amount of crack growth is multiplied by a correction factor to allow for the retardation, if any, due to prior higher loads. The correction factor C_p is given by

$$C_p = \left(\frac{r_{pi}}{r_{p0}}\right)^m \qquad (6.4)$$

where, in Wheeler's notation, m is an empirically adjusted shaping exponent, r_{pi} is the plastic zone size (Section 4.4) calculated from K_{max} for the ith cycle being considered, and r_{p0} is the distance from the current crack tip to the greatest extent reached by a similarly calculated plastic zone for a previous load, i.e. the distance to the elastic/plastic boundary.

The situation is further complicated when individual cycles cannot be distinguished. Various methods of calculating equivalent cyclic spectra have been developed,[19] but all have shortcomings; the most logically defensible and sophisticated is the 'rainflow' method, in which rain is imagined to flow down a load/time record plotted vertically (Fig. 6.11). Flow starts at the beginning of the record, then the inside of each peak in the order in which peaks are applied. It stops when it either meets flow from a higher level, or a point opposite a peak which is arithmetically greater (or equal to) the point from which it started, or when the end of the record is reached. Each separate flow is counted as a half-cycle. There is always a complementary half-cycle of opposite sign, except perhaps for a flow which either starts at the beginning of the record or reaches the end. This can be minimized by using a development known as the reservoir method.[30]

The many variables make it difficult to predict the fatigue lives of structures under variable amplitude loading, even where the predictions are for tests carried out under carefully controlled laboratory conditions. For example, attempts have been made to forecast the lives of three sets of welded joints,[31] tested under narrow-band random loading, with the results of the tests being unknown at the time the forecasts were made. None of the forecasts was particularly satisfactory, although once the results of the tests were known it was always possible to make plausible assumptions which led to an improved prediction.

The advent of servohydraulic fatigue testing equipment has made it easy to apply a complex load history to a structure or test specimen

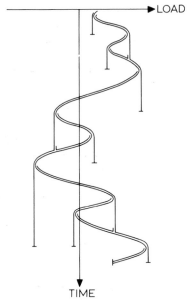

LOAD

TIME

Fig. 6.11   The rainflow method of cycle counting (from Chell[10])

(Sections 2.5 and 3.3.2), and this has considerably simplified the determination of life under variable amplitude loading. However, in structures where safety is critical and testing of representative structures is impractical, such as nuclear pressure vessels, elaborate theoretical calculations may be carried out.[32]

REFERENCES

1   W. ELBER: *Eng. Fract. Mech.*, 1970, **2**, (1), 37–45
2   N. E. FROST: *J. Mech. Eng. Sci.*, 1962, **4**, (1), 22–35
3   M. C. LAFARIE-FRENOT, J. PETIT and C. GASC: *Fatigue Eng. Mater. Struct.*, 1979, **1**, (4), 431–438
4   D. L. DAVIDSON and J. LANKFORD: *ibid.*, 1979, **1**, (4), 439–446
5   M. NAKAGAKI and S. N. ATLURI: *ibid.*, 1979, **1**, (4), 421–429
6   N. E. FROST, K. J. MARSH and L. P. POOK: 'Metal fatigue', 1974, Oxford, Clarendon Press
7   J. M. KRAFFT, A. M. SULLIVAN and R. W. BOYLE: Crack Propagation Symposium, Vol. 1, 8–28, 1961, Cranfield, College of Aeronautics
8   L. P. POOK and N. E. FROST: *Int. J. Fract.*, 1973. **9**, (1), 53–61

9 L. P. POOK and A. F. GREENAN: Int. Conf. on Fatigue Testing and Design, City University, London, 1976; Vol. II, 30.1–30.33, 1976, Buntingford, Herts., Society of Environmental Engineers, Fatigue Group

10 G. G. CHELL (ed.): 'Developments in fracture mechanics—I', 1979, London, Applied Science Publishers Ltd

11 C. BATHIAS and J.-P. BAÏLLON (eds.): 'Le fatigue des matériaux et des structures' ('Fatigue of materials and structures'), 1980, Paris, Maloine SA

12 S. J. HUDAK, A. SAXENA, R. J. BUCCI and R. C. MALCOLM: Report No. AFML-TR-78-40, 1978, Wright–Patterson Air Force Base, Ohio, Air Force Materials Laboratory

13 S. PEARSON: Eng. Fract. Mech., 1975, 7, (2), 235–247

14 N. E. FROST, L. P. POOK and K. DENTON: ibid., 1971, 3, (2), 109–126

15 P. J. E. FORSYTH: J. Soc. Environ. Eng., 1980, 19-1, (84), 3–10

16 L. P. POOK: 'Residual stress effects in fatigue', ASTM STP 776, 97–114, 1982, Philadelphia, Pa., American Society for Testing and Materials

17 D. P. ROOKE and D. J. CARTWRIGHT: 'A compendium of stress intensity factors', 1976, London, HMSO

18 'Fatigue crack growth under spectrum loads', ASTM STP 595, 1976, Philadelphia, Pa., American Society for Testing and Materials

19 P. WATSON and B. J. DABELL: J. Soc. Environ. Eng., 1976, 15-3, (20), 3–9

20 W. SCHÜTZ: Eng. Fract. Mech., 1979, 11, (2), 405–421

21 J. C. NEWMAN: 'Design of fatigue and fracture resistant structures' (eds. P. M. Abelkis and C. M. Hudson), ASTM STP 761, 255–277, 1982, Philadelphia, Pa., American Society for Testing and Materials

22 J. SCHIJVE: Eng. Fract. Mech., 1974, 6, (2), 245–252

23 L. P. POOK: Weld. Res. Int., 1974, 4, (3), 1–24

24 T. R. BRUSSAT: 'Damage tolerance in aircraft structures', ASTM STP 486, 122–147, 1971, Philadelphia, Pa., American Society for Testing and Materials

25 J. N. EASTABROOK: Tech. Rep. 81029, 1981, Farnborough, Procurement Executive, Ministry of Defence

26 A. E. CARDEN, A. J. McEVILY and C. H. WELLS (eds.): 'Fatigue at elevated temperature', ASTM STP 520, 1973, Philadelphia, Pa., American Society for Testing and Materials

27 L. P. POOK and A. F. GREENAN: Int. J. Fatigue, 1979, 1, (1), 17–22

28 D. BROEK and S. H. SMITH: Eng. Fract. Mech., 1979, 11, (1), 123–141

29 O. E. WHEELER: J. Bas. Eng., 1972, 94D, (1), 181–186

30 BRITISH STANDARDS INSTITUTION: 'Steel, concrete and composite bridges. Part 10: Code of practice for fatigue', BS 5400: Part 10: 1980

31 L. P. POOK and R. HOLMES: 'Advances in fracture research, Vol. 5' (ed. D. François), 2079–2091, 1981, Oxford, Pergamon Press

32 P. C. RICCARDELLA and T. R. MAGER: 'Stress analysis and growth of cracks', ASTM STP 513, 260–279, 1972, Philadelphia, Pa., American Society for Testing and Materials

CHAPTER 7

Fatigue crack growth direction

Let there be some more test made of my metal.
Shakespeare, *Measure for Measure*, I.i.48.

7.1 INTRODUCTION

Several topics have been collected together in this chapter. All involve consideration of the direction taken by a fatigue crack growing in an isotropic material under essentially elastic conditions, so that the crack tip stress field can be characterized by stress intensity factors. It is assumed, as is usually the case, that on a macroscopic scale the crack plane and direction are intuitively obvious.

Discussion is on the basis (Section 4.2) that cracks in an isotropic, elastic material tend to grow in Mode I (Fig. 4.3). This can at best be regarded as a useful generalization based on observation. It does not appear to be susceptible to proof in any strict sense[1] of the word. It appears in the literature in a number of forms, usually without justification, although the occurrence of Mode I crack growth could equally well be used, assuming elastic behaviour, as a test of isotropy or, assuming isotropy, as a test of essentially elastic behaviour.[2] For example, in Ref. 3 it appears as a self-evident axiom for a perfectly isotropic material, with a statistical argument to justify its use in practical situations where crack direction and isotropy can only be defined in a macroscopic sense. An equivalent form is the criterion of local symmetry.[4] This takes as self-evident that a crack tends to grow such that the crack tip stress field becomes symmetrical, and has been used as the basis of recent theoretical analysis of crack direction.[5] Attempts at justification based on thermodynamic criteria neglect the point that satisfaction of an appropriate thermodynamic criterion is a necessary, but not a sufficient, condition for crack growth.[6]

7.2 SLANT CRACK GROWTH IN THIN SHEETS

The transition from square (90°) to slant (45°) crack growth sometimes observed in thin sheets (Sections 4.2 and 4.5) is an exception to the generalization that crack growth tends to take place in Mode

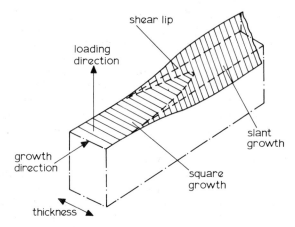

Fig. 7.1 Fracture surface of a fatigue crack with the transition to slant crack growth (from Schijve[7])

I. The major features of the transition[7] are shown in Fig. 7.1. After the transition, crack growth is in a mixture of Modes I and III, but in calculations it is conventional to treat it as if it were Mode I (Section 4.5). The transition to slant crack growth originates in the development of shear lips which increase in width until they reach a maximum size; if the specimen is sufficiently thin they meet, completing the transition. For a given set of circumstances it is usually possible[8,9] to correlate the appearance of shear lips with the attainment of a critical value of ΔK_I (or rate of crack growth), and the completion of the transition with a somewhat higher critical value.[7] The length of the transition zone of mixed square and slant fracture therefore depends on the rate of increase of ΔK_I with crack length. It is possible to reverse the transition by reducing the load level.[10] Whether or not the transition takes place can depend on specimen geometry.[11] The only generalizations that can be made with reasonable confidence[9] are that the attainment of a critical value of ΔK_I is a necessary condition for the appearance of shear lips, and a state of plane stress at the maximum load in the fatigue cycle is a necessary condition for completion of the transition, but neither condition is by itself sufficient.

The mechanism responsible for the transition to slant crack growth in thin sheets is not clear, although the actual crack growth mechanisms are the same as in Mode I fatigue crack growth.[9] The problem of what causes slant growth reduces to finding the conditions in which a shear lip will develop. Shear lips are associated with the plane stress conditions which exist at the surfaces even of thick specimens.[9] The appearance of shear lips in thick specimens rules out buckling as a cause. Shear lips appear on both sides of a specimen at about the same time, so their appearance must be due to the satisfaction of some condition. In thick specimens, shear lips on opposite sides of a specimen appear randomly on planes parallel and perpendicular to each other, whereas in thin specimens they usually, but not invariably, appear on parallel planes (Fig. 7.2).

A possible explanation of the transition, which has the advantage of conceptual simplicity,[9] is that the development of shear lips is an instability effect. The problem then reduces to a question of what happens when a small random deviation in the plane of crack growth occurs at a specimen surface. If conditions are such that further growth is towards the original plane, then the situation is stable and there is no transition. If further growth deviates further from the original plane the situation is unstable and a shear lip develops. A somewhat analogous, but essentially two-dimensional, situation exists when crack directional stability is considered (Section 7.5).

Conventionally (Section 4.5), no distinction is made between square and slant growth in the presentation of fatigue crack growth data. Evidence on whether crack growth rates decrease or increase after the transition is conflicting.[8] Examination of the model described in Section 4.9 gives some guidance. If it is assumed that the sequence of crack opening and closing remains as shown in Fig. 2.4, then the

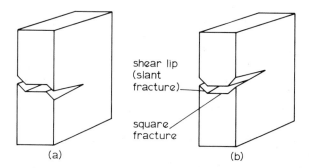

shear lip
(slant
fracture)

square
fracture

(a)   (b)

Fig. 7.2   Shear lips on (a) parallel and (b) perpendicular planes (from Pook[9])

model[12] leads to*

$$\frac{da}{dN} = \frac{9 \cdot 6}{\pi} \left(\frac{\Delta K_{\mathrm{I}}}{E} \right)^2 \tag{7.1}$$

i.e. the crack growth rate increases slightly after the transition, which is what might be expected on thermodynamic grounds. This assumes implicitly that the fracture surface is 'rough', so that Mode III displacements are prevented as soon as the crack surfaces meet. If the fracture surface is 'smooth', then Mode III displacements are not prevented, and only Mode I displacements contribute to crack growth, leading to

$$\frac{da}{dN} = \frac{3 \cdot 1}{\pi} \left(\frac{\Delta K_{\mathrm{I}}}{E} \right)^2 \tag{7.2}$$

i.e. the crack growth rate decreases after the transition.

It is clear that both the start of the transition and crack growth rates after the transition depend on the fracture morphology. It is difficult to proceed further without precise definition (Section 5.5.1) of what is meant by 'rough' and 'smooth' in various contexts. However, it is interesting to note that the suppression of shear lip formation in aluminium alloys in aggressive environments[13] is associated with fracture surfaces having a smooth appearance.

7.3 MIXED MODE THRESHOLD BEHAVIOUR

Conventional specimens used to determine crack growth properties of materials are designed so that only Mode I displacements are present. However, in the general case of a crack-like flaw from which a service failure originates, Mode II or Mode III displacements or both may be present. Under such mixed mode loadings, initial crack growth is generally not in the plane of the initial flaw, and stress intensity factors change radically as soon as crack growth starts. The direction of crack growth is normally such that K_{I}, the opening mode stress intensity factor, has its maximum value and K_{II} and K_{III}, the edge-sliding and shear mode stress intensity factors, are zero. As cracks tend to grow in Mode I, it is pertinent to consider the behaviour of a small Mode I branch at an initial (main) crack. Short-crack limitations (Section 4.8) mean that stress intensity factors are not a valid basis for the discussion of mixed mode problems unless the initial (main) crack is at least of the order of $0 \cdot 25$ mm long.[2]

---

* The factor given in Ref. 12 is incorrect.

In this section, the prefix Δ is only added where it is needed for clarity, and the following conditions are assumed. The threshold is sharply defined, although this is not always the case (Sections 4.7, 6.2 and 6.4). Except where noted, K_{II} and K_{III}, which can be of either sign, do not pass through zero during the fatigue cycle. Ratios between K_I, K_{II} and K_{III} do not vary during loading. (An example of non-proportional fatigue loading is discussed in Ref. 14.) The crack is initially stress free; in particular, there are no residual stresses due to a crack tip plastic zone developed by some type of prior loading. The main crack is straight and, for three-dimensional cracks, the main crack is flat and its crack front straight, or the appropriate radii of curvature are large compared with branch crack length.

Consider first the general case of an angled main crack, where only Mode I and II displacements are present. An angled crack is usually shown[2] as a central crack in a large sheet (Fig. 7.3) with the crack inclined to the applied stress. Here, it is taken to mean the quasi-two-dimensional case where only Mode I or Mode II displacements, or a combination of the two, can be present. For an angled crack, crack growth is at an angle θ to the initial (main) crack, as shown schematically[2] in Fig. 7.4; for θ to be positive, K_{II} must be negative. Criteria are needed for the formation of a branch crack, its initial direction and whether or not the branch, once formed, will continue to grow. Crack growth from an angled crack follows a curved path.[8] A complete solution of the problem would require determination of this path and corresponding stress intensity factors. Here it is assumed that, once a branch starts to propagate, failure

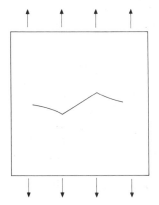

Fig. 7.3   Crack growth from an angled crack (from Pook[2])

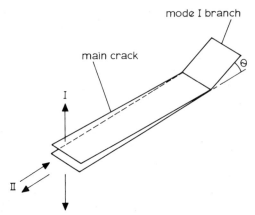

Fig. 7.4   Quasi-two-dimensional crack with Mode I branch (from Pook[2])

will follow. It is important to distinguish between criteria for the formation of a branch crack and criteria for its propagation. Depending on circumstances, either event may dominate behaviour.[2]

Examination of the behaviour of a branch crack is appropriate where the branch forms easily, or can be regarded as already present, perhaps due to a metallurgical discontinuity, so that branch crack propagation is the controlling event. As cracks tend to grow in Mode I, values of θ are selected such that K_{II} for the branch crack is zero, and K_I for the branch crack (k_I) has its maximum value k_I^*. The value of k_I in the vicinity of its maximum is only weakly dependent on θ. The failure envelope obtained from this approach[2,14] is shown in Fig. 7.5 (curve for $K_{III}/k_I^* = 0$).

In practical angled-crack testing, variations from the ideal quasi-two-dimensional crack shape shown in Fig. 7.4 can present problems. In pure Mode I, variations from the ideal crack shape merely lead to some uncertainty in the value of K_I, but in mixed mode testing such deviations can introduce an unwanted mode. For example,[15] in a specimen intended to give pure Mode II displacements, crack front curvature introduces unwanted Mode III displacements and also means that the branch crack cannot be flat. In angled-crack testing in fatigue, extraneous Mode III displacements appear to facilitate branch crack formation, and threshold values are close to those predicted by the failure envelope.[15,16] In the absence of Mode III displacements, branch crack formation, and hence threshold

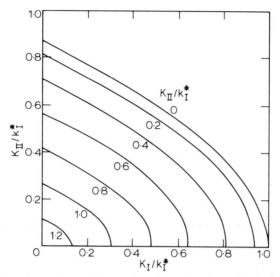

Fig. 7.5   Failure envelope for mixed Mode I, II and III loading (from Pook[2])

behaviour, appears to be largely controlled by the value of K_I for the main crack, with the value of K_{II} having little effect. For pure Mode II, failure takes place away from the crack tip, either by fretting (Section 7.7) 2–3 mm from the crack tip, or at a stress concentration, giving an apparent threshold up to twice that predicted by the failure envelope.[15,16]

Understanding the behaviour of a Mode III crack requires determination of the stress intensity factor for a Mode I branch at a Mode III crack. This is more difficult than the angled-crack problem because of its three-dimensional nature. An approximate solution[2] predicts that the Mode III threshold should be about 1.35 times the Mode I value, compared with an observed[17] ratio for mild steel of $1\frac{1}{4}$. Because a preferred plane of crack growth only intersects the main crack at one point, the subsequent pattern of crack growth is complex[17] (Fig. 7.6).

Extension to the general case of mixed Mode I, II and III loading is straightforward; Fig. 7.5 shows the failure envelope obtained. As it gives the condition for the propagation of a Mode I branch crack, but not its initiation, it is a lower bound. Limited data[16] for mixed Mode I and III loading are close to the lower bound.

Fig. 7.6 Typical fracture of a sharply notched torsion specimen (from Pook and Sharples[17])

7.4 TRANSITION FROM STAGE I TO STAGE II FATIGUE CRACK GROWTH

A fatigue crack originates (Section 2.3) on a crystallographic plane lying in a plane of maximum shear stress so that, depending on the shape of the crack, Mode II or Mode III displacements, or both (Fig. 4.2), will be present. For uniaxial loadings, Mode I displacements will also be present, but not in pure shear (e.g. torsion) loadings. The early stages of crack growth (Stage I in Forsyth's[18] notation) are on the same plane, but at some critical stage the crack changes direction and grows such that only Mode I displacements are present (Stage II crack growth). Stage I crack growth, sometimes called microcrack growth, is therefore an exception to the generalization that fatigue cracks tend to grow in Mode I. Geometric incompatibility can delay the changeover if, as in pure shear, a plane for Mode I crack growth only intersects the Stage I plane at one point.[8] Under appropriate conditions, it is possible to reverse the transition.[19]

It is implicit in the accepted mechanism of fatigue crack initiation (Section 2.3) that a grain in which initiation takes place must have yielded to some extent. When extensive yielding takes place, even long cracks do not necessarily grow in Mode I. For example, Mode II crack growth, in the same plane as the initial crack, has been observed in angled-crack specimens (Fig. 7.3), with the initial crack inclined at 45° so that it is on a plane of maximum shear stress in the uncracked specimens.[20] Such crack growth is perhaps best regarded as an extended form of Stage I growth. There is virtually no information available on crack growth rates during Stage I; some data have recently been obtained for long cracks growing in pure Mode III.[21,22]

The basic problem is the determination of the conditions under which the crack changes direction. Clearly, the transition to Stage II crack growth cannot take place unless an appropriate mixed mode threshold criterion (see previous section) is satisfied. The mechanism of Stage I growth is unclear, but extensive yielding does appear to be required. In Mode I, a fatigue crack grows on each cycle by a deformation mechanism involving the opening and closing of the crack tip (Section 4.9). The argument can be extended to Mode III displacements but not to Mode II.[12] Mode II crack growth at zero mean load can, however, be explained if it is assumed that the crack changes direction on each half-cycle, as shown schematically in Fig. 7.7a. A detailed analysis is not possible because of interference between the crack surfaces. The argument may be extended to cover unidirectional loadings by noting that a yielded grain in which a

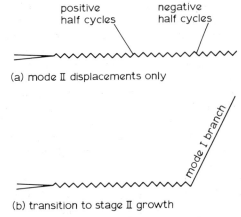

positive
half cycles

negative
half cycles

(a) mode II displacements only

mode I branch

(b) transition to stage II growth

**Fig. 7.7   Schematic of Stage I fatigue crack growth at zero
mean stress (from Pook and Greenan[20])**

fatigue crack initiates is loaded under constant strain, rather than
constant stress, and that under constant strain loading the mean
stress tends to zero when yielding takes place.[23]

If a branch became established (Fig. 7.7b) such that only Mode I
displacement were present, perhaps as a result of a metallurgical
discontinuity, then crack growth would be expected to continue along
this branch. This suggests that the transition to Stage II crack growth
is controlled by the directional stability of a Stage I crack, not solely
by threshold phenomena. A crack is likely to change direction when
it meets a metallurgical discontinuity such as a grain boundary or
inclusion.

7.5 BIAXIAL STRESSES

The stress intensity factor (Section 4.3) is actually the first term in
the series expansion for a crack tip stress field;[24] it dominates stresses
at the crack tip and therefore controls crack propagation rates. Its
value is unaffected by applied stresses parallel to the crack, so the
degree of biaxiality has little influence on crack propagation rates.
The second term (see Appendix I) represents a stress T parallel to
the crack and is therefore strongly dependent on biaxiality. For static
loading, an elastic analysis[25] (which can be modified for crack tip
plasticity) shows that it controls crack directional stability. Provided
that T is negative, as it is for most fracture mechanics test speci-

mens,[26] the crack is directionally stable: after a small random deviation it tends to return to its original direction. When T is positive, as it is for a small central crack in a sheet with an applied tensile stress parallel to the crack greater than the applied tensile stress perpendicular to the crack (a biaxiality ratio greater than 1), the crack is directionally unstable, and the crack trajectory becomes a random walk whose precise path cannot be predicted, even in principle, with certainty.[25] Figure 7.8 shows the typical S-shaped trajectories[27] obtained under static biaxial loads; the crack turns increasingly sharply as the biaxiality ratio increases. Behaviour under fatigue loads is generally similar,[28,29] but in detail is strongly material dependent. Data are very sparse.

After a test is complete, it is possible to demonstrate by computer analysis (e.g. Ref. 30) that a particular curved fatigue crack has been growing in Mode I. Theoretical predictions of crack trajectories appear to be limited to certain restricted cases of static loading under essentially elastic conditions.[5,25] Cracks in structures often follow complex paths (e.g. Fig. 7.9), and at present the only way to accumu-

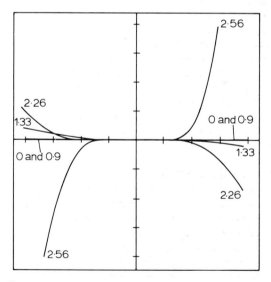

Fig. 7.8 Fracture trajectories under biaxial stress in 6.35 mm thick specimens. Numbers indicate biaxiality ratios (from Chell[27])

Fig. 7.9 Fatigue failure from a spot weld in a propeller shaft

late information that can be applied to design is through appropriate structural tests.

7.6 CRACK SHAPE

Most conventional fracture mechanics specimens may be regarded as essentially two-dimensional, with the stress intensity factor constant along the crack front. In general, this is not true for structures; the crack front is usually curved with the stress intensity factor, and hence fatigue crack growth rates, varying along the crack front.[31] The so-called part-through crack, which is a surface crack of roughly semi-elliptical profile (Figs 2.3 and 7.10), is of particular practical importance.[31,32]

In principle, the change in crack shape as a crack grows can be predicted by calculating stress intensity factors and hence crack growth increments at various points along the crack front. In practice, the rapid accumulation of errors as successive increments are considered makes such predictions unreliable, as has been shown experimentally.[32] However, cracks do tend towards a stable shape that is largely independent of the initial shape, but depends on specimen geometry and loading conditions.[33,34] As the relative stress intensity factor distribution does not depend on the absolute value of the load level, it might appear that crack shape development would be largely independent of the absolute load level. However, in practice, the crack shape may also change with changing absolute load level, as shown in Fig. 7.10 for an aluminium alloy[32] where crack profiles at constant depth are compared. Also included is the profile for stable growth under a static load.

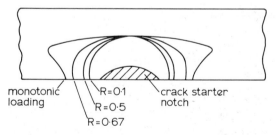

Fig. 7.10 Effect of the R-ratio on crack shape in an aluminium alloy at constant alternating load and crack depth (from Hodulak, Kordisch, Kunzelmann and Sommer[32])

7.7 FRETTING FATIGUE

When two pieces of metal are pressed together, usually by an external static load, and are subjected to a cyclic loading such that one contacting face is cyclically displaced tangentially to the other face, wear of the mating surfaces occurs. If the magnitude of the slip displacement does not exceed about 10^{-1} mm, the wear is termed fretting.[8] Fretting occurs by contacting asperities on the mating surfaces continually welding together and then breaking. This leads to surface pitting and the transfer of metal from one surface to another. In addition, the small fragments of metal which are broken off oxidize, forming oxide particles which for most practical metals are harder than the metal itself. These become trapped between the mating surfaces and cause abrasive wear and scoring. Thus, in certain applications, fretting can lead to a loss of fit between the two mating parts. Examples of situations where fretting may occur in service are in a tapered cone and shaft assembly, in a press fit wheel and shaft assembly, in pin, bolted or riveted joints, and between leaf springs. Unusually, fretting has been observed on cracks loaded in Mode II (Section 7.3); this is due to interference between irregularities on the crack surface rather than to the external load.

Fretting during fatigue loading can produce large reductions in fatigue strength. However, those fretting conditions which give a high rate of surface damage (large slip amplitudes and high contact pressures) do not necessarily produce the greatest loss of fatigue strength. The lowest fatigue strength is usually associated with a slip amplitude of around 10^{-2} mm.[8,35] Large numbers of fretting fatigue tests have been carried out under controlled conditions, and much practical information on the alleviation of fretting fatigue is available.[8,35,36] A typical fretting fatigue test[36] is shown in Fig. 7.11. Narrow pads are machined at either end of blocks not necessarily of the same material as the specimens. Using a calibrated steel ring, the pads are clamped against flats machined along a conventional, direct stress, plain fatigue specimen. The origin of the fatigue fracture can be seen beneath the lower pad.

Fatigue damage due to fretting takes the form of surface cracks which penetrate the fretted material.[36] Thus, the fatigue behaviour of a fretted part is largely a matter of crack propagation; some recent work on fretting fatigue has been devoted to the determination of relevant stress intensity factors.[35] Under some conditions, the fretting fatigue strengths of mild steel and alloy steel are about the same;[8] this is because the fatigue crack growth properties are similar (Section 4.5.1).

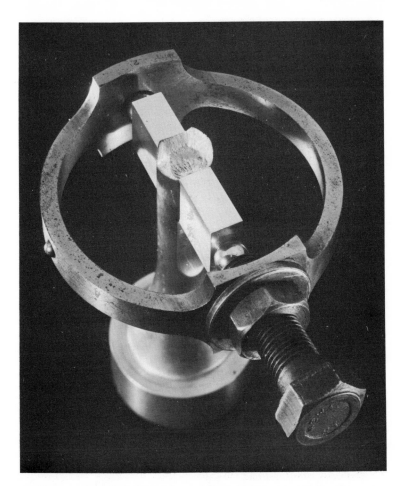

Fig. 7.11 Typical fretting fatigue test (from Field and Waters[36])

The mechanism of fretting fatigue appears to be as follows.[8,36] Under normal contact load and cyclic tangential forces, the oxide film on the surface of contacting asperities breaks down and, with some metals, very strong metal-to-metal adhesion occurs. If the cyclic tangential slip displacements are insufficient to rip apart the micro-weld, the welded junction may spread over a relatively large area, forming a miniature spot weld and making a fracture mechanics analysis appropriate,[16,17] although in practice there are too many uncertainties for a detailed analysis. Under the loading conditions, only Modes II and III crack surface displacements can be present, so it is a mixed mode problem (Section 7.3).

The microwelds are typically a few hundredths of a millimetre across and first develop after some thousands of fretting cycles. If the average shear stress on the spot weld is above the shear yield stress (which will normally have been elevated by work hardening during fretting), the weld fails and fretting fatigue will not occur. Under essentially elastic conditions, behaviour depends on the formation and propagation of Mode I branch cracks (Section 7.3); and detailed understanding depends on a knowledge of the relevant stress intensity factors. Adjacent branch cracks often run together to release metal particles, leaving small pits; these, however, do not represent as serious a form of fatigue damage as the initial cracks, which often extend beyond the root of the pit at the time of its formation.

REFERENCES

1   A. J. AYER: 'The problem of knowledge', 1956, Harmondsworth, Penguin Books

2   L. P. POOK: NEL Report No. 667, 1980, East Kilbride, Glasgow, National Engineering Laboratory

3   B. COTTERELL: *Int. J. Fract. Mech.*, 1965, **1**, (2), 96–103.

4   R. V. GOLDSTEIN and R. L. SALGANIK: *Int. J. Fract.*, 1974, **10**, (4), 507–523

5   B. COTTERELL and J. R. RICE: *ibid.*, 1980, **16**, (2), 155–169

6   A. H. COTTRELL: 'The mechanical properties of matter', 1964, London, John Wiley & Sons

7   J. SCHIJVE: *Eng. Fract. Mech.*, 1981, **14**, (4), 789–800

8   N. E. FROST, K. J. MARSH and L. P. POOK: 'Metal fatigue', 1974, Oxford, Clarendon Press

9   L. P. POOK: *Met. Sci.*, 1976, **10**, (9), 334–335

10   J. SCHIJVE: *Eng. Fract. Mech.*, 1974, **6**, (2), 245–252

11   R. O. RITCHIE, R. F. SMITH and J. F. KNOTT: *Met. Sci.*, 1975, **9**, (11), 485–492

12   L. P. POOK and N. E. FROST: *Int. J. Fract.*, 1973, **9**, (1), 53–61

13   L. B. VOGELESANG and J. SCHIJVE: *Fatigue Eng. Mater. Struct.*, 1980, **3**, (1), 85–98

14   L. P. POOK: 'Fracture and fatigue. Elastoplasticity thin sheet and micromechanism problems' (ed. J. C. Radon), 143–153, 1980, Oxford, Pergamon Press

15   L. P. POOK and A. F. GREENAN: 'Fracture mechanics' (ed. C. W. Smith), ASTM STP 677, 23–25, 1979, Philadelphia, Pa., American Society for Testing and Materials

16   L. P. POOK: 'Fatigue thresholds. Fundamentals and engineering applications' (eds. J. Bäcklund, A. F. Blom and C. J. Beevers), 1007–1032, 1982, Warley, West Midlands, Engineering Materials Advisory Services Ltd

17   L. P. POOK and J. K. SHARPLES: *Int. J. Fract.*, 1979, **15**, (6), R223–R225

18   P. J. E. FORSYTH: 'The physical basis of metal fatigue', 1969, London, Blackie and Son Ltd

19   M. W. BROWN and K. J. MILLER: *Fatigue. Eng. Mater. Struct.*, 1979, **1**, (2), 231–246

20   L. P. POOK and A. F. GREENAN: Int. Conf. on Fatigue Testing and Design, City University, London, 1976; Vol. II, 30.1–30.33, 1976, Buntingford, Herts., Society of Environmental Engineers. Fatigue Group

21   P. E. IRVING and N. J. HURD: 'Design of fatigue and fracture resistant structures' (eds. P. R. Abelkis and C. M. Hudson), ASTM STP 761, 212–233, 1982, Philadelphia, Pa., American Society for Testing and Materials

22   R. O. RITCHIE, F. A. McCLINTOCK, H. NAYEB-HASHEMI and M. A. RITTER: *Metall. Trans. A*, 1982, **13A**, (1), 101–110

23   B. I. SANDOR: 'Fundamentals of cyclic stress and strain', 1972, Madison, Wis., University of Wisconsin Press

24   M. L. WILLIAMS: *J. Appl. Mech.*, 1957, **24**, (1), 109–114

25   P. S. LEEVERS and J. C. RADON: *J. Mech. Phys. Solids*, 1976, **24**, (6), 381–395

26   P. S. LEEVERS and J. C. RADON: *Int. J. Fract.*, 1982, **19**, (4), 311–325

27   G. G. CHELL (ed.): 'Developments in fracture mechanics—I', 1979, London, Applied Science Publishers Ltd

28   L. P. POOK and R. HOLMES: Int. Conf. on Fatigue Testing and Design, City University, London, 1976; Vol. II, 36.1–36.22, 1976, Buntingford, Herts., Society of Environmental Engineers, Fatigue Group

29   A. F. LIU, J. E. ALLISON, D. F. DITTMER and J. A. YAMANE: as Ref. 15, 5–22

30   S. IIDA and A. S. KOBAYASHI: *J. Bas. Eng.*, 1969, **91**, (4), 764–769

31   S. J. HOLDBROOK and W. D. DOVER: *Eng. Fract. Mech.*, 1979, **12**, (3), 347–364

32   L. HODULAK, H. KORDISCH, S. KUNZELMANN and E. SOMMER: *Int. J. Fract.*, 1978, **14**, (2), R35–R38

33   L. HODULAK, H. KORDISCH, S. KUNZELMANN and R. SOM-
     MER: as Ref. 15, 399–410
34   W. S. PIERCE and J. C. SHANNON: *J. Test. Eval.*, 1978, **6**, (3),
     183–188
35   R. B. WATERHOUSE (ed.): 'Fretting fatigue', 1981, London, Applied
     Science Publishers Ltd
36   J. E. FIELD and D. M. WATERS: NEL Report No. 275, 1967, East
     Kilbride, Glasgow, National Engineering Laboratory
37   L. P. POOK: *Int. J. Fract.*, 1975, **11**, (1), 173–176

CHAPTER 8

Metal fatigue today

We have left undone those things which we ought
to have done; And we have done those things
which we ought not to have done.
The Book of Common Prayer, *Morning Prayer*.

8.1 INTRODUCTION

The aim of this closing chapter is briefly to describe where the subject
of metal fatigue now stands. As mentioned in Chapter 1, much has
been written on the problem of metal fatigue since 1838. One might
argue from the infrequency of catastrophic fatigue failures in
engineering structures that the problem is no longer serious. Clear
explanations are usually found for the major failures that do occur,
with human error, it seems, often being a culprit. However, lesser
fatigue failures, often unrecognized unless they happen to be seen
by a fatigue specialist, are a common and expensive nuisance.

A UK Government Report on the fatigue of engineering structures,[1] issued in 1960, urged the provision of funds and facilities so
that fatigue information appropriate to engineering design could be
generated on a much larger scale. In the intervening years the facilities
have appeared, as they have in many other countries, and a vast
amount of data has been acquired. General awareness of the dangers
of metal fatigue has greatly increased, and by and large the problem
can be contained, if not solved; but the price is eternal vigilance.
However, the subject is not mature in the sense that none of the
books on metal fatigue can be regarded as a standard text.

In attempting to keep metal fatigue in perspective, it must always
be remembered that fatigue is only one of many, often conflicting
requirements with which a designer must contend. Innovation always
brings the risk of fatigue failure, but because fatigue is not a suddenly
occurring phenomenon, and owing to the need to pursue detail, it
is only possible to progress slowly as experience is acquired in fresh
areas. Again, because of the need to pursue detail, the transfer of
metal fatigue technology from producer to user tends to be a frustrat-
ing and expensive process. For instance, a single symposium held in

the United States in 1978 was estimated[2] to involve total costs of the order of 1 000 000 dollars.

Metal fatigue is very much a descriptive subject, with a voluminous literature. Metallurgical descriptions are concerned with the effect of fatigue loading on the state of the material. Mechanical descriptions are concerned with matters such as the number of cycles to failure, or the rate of growth of a fatigue crack, and are the more useful from an engineering viewpoint.

8.2   AVOIDANCE OF FATIGUE FAILURE

Even rough estimation of fatigue strength at the preliminary design stage presents problems. At first sight this is surprising, since large amounts of experimental data are available and, in general, the basic mechanisms of fatigue failure are relatively well understood and documented. Problems in estimating fatigue strength arise because of the complexity and variety of realistic engineering design situations. This makes it difficult to generalize on known fatigue data and then apply these generalizations to a specific engineering problem. In the past, most advances in design against fatigue were made on a trial-and-error basis.

In principle, an analytical approach, based on materials data and applied mechanics, can be used to predict the service life of a structure. The crack initiation and crack propagation phases must be treated separately, although in practice one or the other usually predominates. An analytical approach requires expert knowledge, and often fails because of a lack of the detailed information required. A particular problem is that there is no simple relationship between the scatter observed in fatigue crack growth rate data and the scatter in the service life of structures containing cracks. Simplified calculations can be useful when an approximate answer is adequate, although experienced judgment is needed in making the necessary simplifications. Simplified fracture mechanics calculations have proved particularly valuable in failure analysis. As new information becomes available, the range of situations which can be analysed is increasing.

Standardized procedures of various degrees of formality are the most satisfactory from the designer's viewpoint. They may be based on analytical procedures, service experience, the results of structural tests or some combination of these, and need not have a theoretically sound basis provided they give sufficiently accurate answers. Very simple methods are often used in the early stages of design, and ideally no expert knowledge is required. These methods fail where no appropriate standard procedure is sufficiently accurate for the situation being considered. The development of standard procedures, which can be a lengthy process, is probably the best way of assimilating the results of research into engineering practice. Most fatigue assessments are based on simplified standardized procedures, which despite their apparent lack of physical validity, are known to give conservative answers.

Service loading testing is often the most cost effective method of determining service life. Modern servohydraulic equipment permits the application of virtually any load history, so service loading testing can be used to determine service life where analytical and standard procedures fail. This kind of testing, already well established in the aircraft industry, is being increasingly used for critical structures in other industries. Such tests are often used to calibrate approximate analytical methods so that these methods can be used to extrapolate the results of mechanical tests to broadly similar situations.

A considerable research and development effort on fatigue is needed whenever the service life of a fundamentally new type of structure needs to be accurately predicted. Long-term background work on fatigue is needed as it is likely to decrease the effort needed for a particular new situation, and also means that a pool of expertise is kept available.

Present trends in the law on product liability are to increase the liability of manufacturers for personal injury and material damage caused by defective products. Under strict liability, it is merely necessary for a plaintiff to prove that a product was defective, although some defences are possible. Legal definitions of 'defective' tend to be based on the performance of the product and, from the designer's point of view, are difficult to reconcile with the essentially random nature of fatigue.

When a structural failure due to fatigue or other causes leads to catastrophe, society as a whole takes an interest through its institutions. In a 'typical' catastrophic fatigue failure, the immediate cause of the failure is usually easily ascertained, but it is more difficult to discover how procedures designed to ensure safety come to fail. Recommendations of official enquiries are often mainly concerned with plugging procedural loopholes.

8.3 RESEARCH

Basic research on fatigue has always had a strong practical bias in that the motivation is usually to acquire data which will help to avoid fatigue failure. In academic circles, it is often assumed that an analytical approach is the ideal to be aimed for. Much of the earlier fatigue work was concerned with the experimental determination of the fatigue strength of plain, or relatively mildly notched, specimens. Numerous variables can affect the fatigue behaviour of a particular material, such as mean stress, surface finish, environment and special situations such as biaxiality of loading. Further complications arise when realistic load histories are introduced. Consequently, no data collection, however large, can be comprehensive in the sense that the need for fatigue testing on an *ad hoc* basis can be eliminated. Fatigue testing is time consuming and expensive, but claimed short-cuts to the acquisition of fatigue data should be treated with caution, as they are usually based on either empirical correlations of limited applicability, or an oversimplified view of the nature of fatigue.

Perhaps the most significant advance in metal fatigue during the past few years is the general realization that most structures contain crack-like flaws which are either introduced during manufacture, especially if welding is used, or form early in service. Virtually the whole life of the structure is occupied by fatigue crack growth from these flaws. An understanding of fatigue crack growth is therefore essential for the understanding and prediction of the fatigue behaviour of many structures. As a result, the experimental determination of fatigue crack growth rates has been emphasized in more recent fatigue work; the use of the fracture mechanics concept of stress intensity factor as a correlating parameter is virtually universal. Again, the numerous variables involved make it impossible to avoid testing on an *ad hoc* basis. Some useful generalizations have emerged; for example, it has been found that fatigue crack growth properties are largely independent of a steel's composition, tensile strength and metallurgical structure. The use of stress intensity factors to analyse crack growth data means that behaviour can be predicted for any cracked body for which a stress intensity factor expression is available; a designer is not limited to situations similar to those used to generate the original data. However, the presence of residual stresses of unknown magnitude can be a serious obstacle to the application of data. Failure to check whether large-scale yielding might be occurring is the commonest error in the use of stress intensity factors, and quoted data for which no check appears to have been made should be regarded as suspect.

In general, it is not possible to predict variable amplitude fatigue crack growth behaviour from the results of constant amplitude tests. Various empirical correlations are possible, and some theoretical progress is being made. The numerous variables make it difficult to predict the fatigue lives of structures under variable amplitude loading even when predictions are for tests carried out under carefully controlled laboratory conditions.

Metal fatigue has long been recognized as a random phenomenom, and the consequent scatter in results complicates both the analysis of experimental data and their subsequent application to engineering problems. Random behaviour occurs because a metallic material is not a homogeneous continuum when viewed on a microscopic scale. Statistical methods of analysing scatter in the results of traditional fatigue experiments involved in the determination of S/N curves are well established. From a statistical viewpoint, a fatigue crack growth test has important differences from a traditional fatigue test, so the established methods of applying statistics to fatigue are not appropriate. The measurements taken in a test do not permit the derivation of adequate statistics to describe the variability of fatigue crack growth, so satisfactory analysis is not possible, and subjective judgments cannot be avoided.

General features of the mechanisms of metal fatigue are well understood, even if details can be complex. Understanding developed surprisingly slowly, with much confused thinking which still persists in some quarters. In particular, the idea that crack growth occurs under fatigue loading not as a consequence of any progressive structural damage, but simply because unloading resharpens the crack tip at each cycle is not always appreciated. Continuing research is revealing new guidelines for the design of fatigue-resistant structures, e.g. the importance of metallurgical structure in the fatigue crack growth behaviour of certain titanium alloys.

The cost of fatigue research is high, but so are the potential savings. In one recent example,[3] application of existing knowledge which enabled a cracked generator rotor to continue in service for two years saved an estimated £13 million.

8.4 THE FUTURE

As an academic discipline, work on metal fatigue within the present paradigm appears to have passed its peak. Much current academic research work on fatigue and related topics is concerned with obsessive pursuit of detail, involving much labour, but yielding little in the way of deeper understanding. Obviously, development work aimed at the resolution of specific practical fatigue problems will continue.

From a practical design viewpoint, it is more important to keep stresses low than to examine fatigue mechanisms. Analysis of service failures is important to avoid similar failures in future, and much advice on fatigue design amounts to hints and tips on how to reduce stresses at potential failure sites. Further codification of existing empirical and experimental data will continue, partly as the result of increasingly strict product liability legislation.

It is of course impossible to predict how and when a shift to a new metal fatigue paradigm might occur, but some clues emerge when the scales at which various aspects are at present discussed are examined. Possible scales range from 10^{-6} mm (ions and electron clouds) to 100 mm (specimen or component). Fatigue crack growth rates of practical interest range from about 10^{-8} to 10^{-2} mm/cycle. Metallurgical descriptions of mechanisms are usually at scales of the same order as metallurgical features, i.e. 10^{-4} to 10^{-2} mm, and are often referred to as micromechanisms; such descriptions are normally qualitative, rather than quantitative, in nature. The applied mechanics framework, which forms the basis of mechanical descriptions (including fracture mechanics), deals largely with macroscopic aspects at scales of 10^{-1} mm and above. Fracture mechanics descriptions of fatigue crack growth at these scales are highly developed. As crack length decreases, the fracture mechanics scales telescope into the structural features scales and mechanical descriptions are less well developed.

Clearly, significant progress could be made if descriptions, particularly quantitative descriptions, were extended to smaller scales. Perhaps the time is ripe for a fresh look at fatigue behaviour at the atomic level, a scale of 10^{-6} mm, in the light of the increased understanding[4] of the theoretical physics of disordered systems. At somewhat larger scales, the recently developed theory of fractals[5] offers the possibility of extending a rigorous mathematical framework to scales of less than 10^{-1} mm.

REFERENCES

1 'Fatigue of engineering structures. Report of the Committee appointed by the Department of Scientific and Industrial Research', 1960, London, Department of Scientific and Industrial Research

2 J. T. FONG (ed.): 'Fatigue mechanisms', ASTM STP 675, 1979, Philadelphia, Pa., American Society for Testing and Materials

3 G. G. CHELL (ed.): 'Developments in Fracture Mechanics—I', 1979, London, Applied Science Publishers Ltd

4 J. M. ZIMAN: 'Models of disorder. The theoretical physics of homogeneously disordered systems', 1979, London, Cambridge University Press

5 B. B. MANDELBROT: 'Fractals: form, chance and dimension', 1977, San Francisco, W. H. Freeman

APPENDIX I

Elastic stress fields

The elastic stress fields and displacements corresponding to the three modes of crack surface displacement are as follows,[1] referring to Fig. 4.3 for notation (where u, v and w are displacements in the x, y, z direction, ν Poisson's ratio and G the shear modulus)

Mode I

$$\sigma_x = \frac{K_I}{(2\pi r)^{1/2}} \cos\frac{\theta}{2}\left(1 - \sin\frac{\theta}{2}\sin\frac{3\theta}{2}\right)$$

$$\sigma_y = \frac{K_I}{(2\pi r)^{1/2}} \cos\frac{\theta}{2}\left(1 + \sin\frac{\theta}{2}\sin\frac{3\theta}{2}\right)$$

$$\tau_{xy} = \frac{K_I}{(2\pi r)^{1/2}} \sin\frac{\theta}{2}\cos\frac{\theta}{2}\cos\frac{3\theta}{2}$$

$$\sigma_z = \nu(\sigma_x + \sigma_y) \qquad\qquad (I.1)$$

$$\tau_{xz} = \tau_{yz} = 0$$

$$u = \frac{K_I}{G}\left(\frac{r}{2\pi}\right)^{1/2}\cos\frac{\theta}{2}\left(1 - 2\nu + \sin^2\frac{\theta}{2}\right)$$

$$v = \frac{K_I}{G}\left(\frac{r}{2\pi}\right)^{1/2}\sin\frac{\theta}{2}\left(2 - 2\nu - \cos^2\frac{\theta}{2}\right)$$

$$w = 0$$

Note that the crack opens into a parabola, and that because a crack is regarded as a mathematical 'cut', θ must lie within the range $\pm\pi$.

Mode II

$$\sigma_x = -\frac{K_{II}}{(2\pi r)^{1/2}} \sin\frac{\theta}{2}\left(2 + \cos\frac{\theta}{2}\cos\frac{3\theta}{2}\right)$$

$$\sigma_y = \frac{K_{II}}{(2\pi r)^{1/2}} \sin\frac{\theta}{2}\cos\frac{\theta}{2}\cos\frac{3\theta}{2}$$

$$\tau_{xy} = \frac{K_{II}}{(2\pi r)^{1/2}} \cos\frac{\theta}{2} \left(1 - \sin\frac{\theta}{2} \sin\frac{3\theta}{2}\right)$$

$$\sigma_z = \nu(\sigma_x + \sigma_y) \tag{I.2}$$

$$\tau_{xz} = \tau_{yz} = 0$$

$$u = \frac{K_{II}}{G} \left(\frac{r}{2\pi}\right)^{1/2} \sin\frac{\theta}{2} \left(2 - 2\nu + \cos^2\frac{\theta}{2}\right)$$

$$v = \frac{K_{II}}{G} \left(\frac{r}{2\pi}\right)^{1/2} \cos\frac{\theta}{2} \left(1 + 2\nu + \sin^2\frac{\theta}{2}\right)$$

$$w = 0$$

Equations (I.1) and (I.2) are written for plane strain; they can be changed to plane stress by writing $\sigma_z = 0$ and substituting $(1-\nu)/(1+\nu)$ for $1-2\nu$ and $2/(1+\nu)$ for $2-2\nu$.

Mode III

$$\tau_{xz} = -\frac{K_{III}}{(2\pi r)^{1/2}} \sin\frac{\theta}{2}$$

$$\tau_{yz} = \frac{K_{III}}{(2\pi r)^{1/2}} \cos\frac{\theta}{2}$$

$$\sigma_x = \sigma_y = \sigma_z = \tau_{xy} = 0 \tag{I.3}$$

$$w = \frac{K_{III}}{G} \left(\frac{2r}{\pi}\right)^{1/2} \sin\frac{\theta}{2}$$

$$u = v = 0$$

Equations (I.1)–(I.3) are for static or slowly moving cracks and were obtained from the series expansion of the stress fields by neglecting higher terms in r. They can be regarded as a good approximation when r is small compared with the other dimensions of the body in the x–y plane and are exact as r tends to zero.

The sign of K_I is taken as positive when the crack surfaces move apart. A negative K_I only has meaning if the crack is regarded as a narrow slit because, if the crack surfaces are pressed together, the crack has no effect on the stress distribution. The signs of K_{II} and K_{III} are conventionally positive when displacements are as shown in Fig. 4.3.

Provided that the notch root radius is small compared with the notch depth, the stresses near the tip of a notch of conic outline are similar in form to those of a crack, but with the origin of r taken at a distance $\rho/2$ outwards from the centre of the notch root, radius ρ

$$\sigma_x \doteq \frac{K_I}{(2\pi r)^{1/2}} \cos\frac{\theta}{2}\left(1-\sin\frac{\theta}{2}\sin\frac{3\theta}{2}\right) - \frac{K_I}{(2\pi r)^{1/2}}\frac{\rho}{2r}\cos\frac{3\theta}{2}$$

$$\sigma_y = \frac{K_I}{(2\pi r)^{1/2}} \cos\frac{\theta}{2}\left(1+\sin\frac{\theta}{2}\sin\frac{3\theta}{2}\right) + \frac{K_I}{(2\pi r)^{1/2}}\frac{\rho}{2r}\cos\frac{3\theta}{2} \qquad \text{(I.4)}$$

$$\tau_{xy} = \frac{K_I}{(2\pi r)^{1/2}} \sin\frac{\theta}{2}\cos\frac{\theta}{2}\cos\frac{3\theta}{2} - \frac{K_I}{(2\pi r)^{1/2}}\frac{\rho}{2r}\sin\frac{3\theta}{2}$$

For a notch which does not close under a given compressive load, a negative value of K_I has meaning. Note also the relationship between stress concentration factors and stress intensity factors

$$K_I = \lim_{\rho \to 0} \frac{\sigma_{\max}}{2}(\pi\rho)^{1/2} \qquad \text{(I.5)}$$

which provides an alternative definition of the stress intensity factor, where $\sigma_{\max}$ is the stress at the root of the notch.

The elastic stress field can be more accurately represented by including further terms.[2] The first four terms for the opening mode, expressed in polar coordinates for compactness, are

$$\sigma_r = \frac{a_1}{4}\left(\frac{2a}{r}\right)^{1/2}\left(5\cos\frac{\theta}{2}-\cos\frac{3\theta}{2}\right) + a_2\cos^2\theta$$

$$+ \frac{a_3}{4}\left(\frac{r}{2a}\right)^{1/2}\left(3\cos\frac{\theta}{2}+5\cos\frac{5\theta}{2}\right)$$

$$+ \frac{a_4}{2}\left(\frac{r}{2a}\right)(\cos\theta+\cos 3\theta)$$

$$\sigma_\theta = \frac{a_1}{4}\left(\frac{2a}{r}\right)^{1/2}\left(3\cos\frac{\theta}{2}+\cos\frac{3\theta}{2}\right) + a_2\sin^2\theta \qquad \text{(I.6)}$$

$$+ \frac{a_3}{3}\left(\frac{r}{2a}\right)^{1/2}\left(5\cos\frac{\theta}{2}-5\cos\frac{5\theta}{2}\right)$$

$$+ 3a_4\left(\frac{r}{2a}\right)(\cos\theta-\cos 3\theta)$$

$$\tau_{r\theta} = \frac{a_1}{4}\left(\frac{2a}{r}\right)^{1/2}\left(\sin\frac{\theta}{2} + \sin\frac{3\theta}{2}\right) - \frac{a_2}{2}\sin 2\theta$$

$$+ \frac{a_3}{4}\left(\frac{r}{2a}\right)^{1/2}\left(\sin\frac{\theta}{2} - \sin\frac{5\theta}{2}\right)$$

$$+ a_4\left(\frac{r}{2a}\right)(\sin\theta - 3\sin 3\theta)$$

The coefficients a_1, a_2, a_3 and a_4 have the dimensions of stress. Because

$$a_1 = \frac{K_I}{2(\pi a)^{1/2}} \tag{I.7}$$

the terms including a_1 are equivalent to the opening mode stress intensity factor K_I. These terms dominate the stress field near the crack tip and control crack initiation; the other terms become increasingly important further away from the crack tip. The second terms control crack directional stability. If the coefficient a_2 is positive, a crack will tend to deviate increasingly from its original direction. If a_2 is negative, a crack will tend to grow in its original direction. The sign of a_2 can be determined from the isochromatic fringes on a photoelastic model; if they lean back, it is positive. The terms in a_2 represent stresses parallel to the crack, and crack deviation can be prevented by applying a compressive load parallel to the crack.

REFERENCES

1 H. TADA, P. C. PARIS and G. R. IRWIN: 'The stress analysis of cracks handbook', 1973, Hellertown, Pa., Del Research Corporation
2 B. COTTERELL: *Int. J. Fract. Mech.*, 1965, **1**, (2), 96–103

APPENDIX II

Determination of S/N curves

To determine the fatigue properties of a material at a given mean stress requires the production of a batch of nominally identical plain specimens; the number in the batch may be limited by the amount of material available, but it is usual for it to be between 6 and 12. If the fatigue properties at zero mean load are required, the specimens can be tested in either a rotating bending (Fig. 2.5) or a direct stress machine; if a mean stress is to be applied, the latter type of machine is used.

The object of testing a batch of specimens is to determine their alternating stress life (stress/number of cycles) relationship so that the maximum alternating stress which will just not cause a specimen to break before a stipulated endurance is achieved can be estimated. If the fatigue properties at zero mean load are required, the first specimen could be tested at an alternating stress of about three-quarters of the tensile strength of the material, for example, $\pm 450 \, \text{MN m}^{-2}$ for a material of $600 \, \text{MN m}^{-2}$ tensile strength. The life to fracture, or endurance (i.e. the number of times the stress cycle is applied), is noted and the next specimen tested at a stress, say, $\pm 30 \, \text{MN m}^{-2}$ lower than that applied in the previous test, and so on with subsequent specimens until a stress level is reached at which the specimen does not break after a stipulated endurance. It is usually found for steels having tensile strengths up to about $700 \, \text{MN m}^{-2}$ that, if a specimen has not broken after 10^7 cycles, it is most unlikely to break at longer endurances; tests on these materials therefore can be terminated at about 20×10^6 cycles. Tests on other materials are often continued up to about 10^8 cycles. The results are usually presented in graphical form, the stress amplitude (or its logarithm) being plotted against the logarithm of the corresponding endurance. Specimens should be tested at sufficient different stress levels to enable a line to be drawn through the experimental points. This line is referred to as an S/N curve; Fig. II.1 shows a typical S/N curve obtained from a batch of carbon steel specimens tested in rotating bending.

The data on Fig. II.1 show that specimens either had a life of less than 5×10^6 cycles or were still unbroken when the tests were stopped

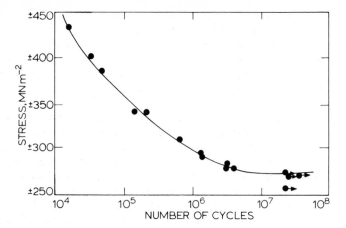

Fig. II.1 *S/N* **curve for carbon steel; specimens tested in rotating bending. Points having arrows attached denote specimen unbroken (from Frost, Marsh and Pook**[1]**)**

at $20\text{--}40 \times 10^6$ cycles and suggest that the line through the points becomes parallel to the abscissa. When this is so, the stress level corresponding to the horizontal line is called the fatigue limit; it is implied that specimens tested at stress levels of less than the fatigue limit will never break no matter how many stress cycles are applied. If the finite life portion of the *S/N* curve can be represented by a straight line, the point of intersection of this line and the horizontal line is often referred to as the 'knee'.

Not all metals give *S/N* curves which exhibit a definite fatigue limit, even when tests are continued to very long endurances. It is usual in this case to specify the fatigue strength of the material at a given endurance, i.e. specimens tested at higher stress levels will break before the stipulated life is achieved, whereas those tested at lower stress levels will be unbroken after a life of at least the stipulated value. However, the slope of an *S/N* curve is generally small at endurances in the region of 10^8 cycles, and it is often possible to make an estimate of a stress level which, for many practical purposes, can be considered as a fatigue limit.

If a number of apparently identical specimens of the same material are tested at the same nominal stress amplitude, it is found that they do not have identical lives, the scatter in lives increasing as the stress level decreases towards the fatigue strength at long endurances. Scatter in results arises from slight variations in testing conditions (for example, exactly the same stress cannot be applied to each specimen) and slight variations in the properties of the surface layers of the material (for example, local differences in microstructure and local surface differences induced by the machining and polishing procedures).

Usually, the number of specimens required to define the S/N curve for a particular material is based on previous experience of the scatter in results known to occur with materials of a similar type. For example, the results of fatigue tests on pure metals, simple alloys (e.g. brass) and steels having tensile strengths up to 700 MN m^{-2} do not show a large scatter, and their S/N curves can generally be defined quite closely by testing about 10 specimens. On the other hand, the results of tests on complex alloys (e.g. high strength steel and aluminium alloys) often exhibit considerable scatter, and at least twice as many specimens may be needed to obtain a reasonably reliable S/N curve.

REFERENCE
1 N. E. FROST, K. J. MARSH and L. P. POOK: 'Metal fatigue', 1974, Oxford, Clarendon Press

Index